STUDENT SOLUTIONS MANUAL

Charles W. Haines
Rochester Institute of Technology

to accompany

Elementary Differential Equations

Fifth Edition

Elementary Differential Equations and Boundary Value Problems

Fifth Edition

William E. Boyce
Richard C. DiPrima
Rensselaer Polytechnic Institute

John Wiley & Sons, Inc.
New York • Chichester • Brisbane • Toronto • Singapore

ISBN 0-471-55127-9

Printed in the United States of America

10 9 8 7 6 5 4 3 2 1

PREFACE

This supplement has been prepared for use in conjunction with the fifth editions of ELEMENTARY DIFFERENTIAL EQUATIONS AND BOUNDARY VALUE PROBLEMS and ELEMENTARY DIFFERENTIAL EQUATIONS, both by W.E. Boyce and R.C. DiPrima. The supplement contains a sampling of problems from each section of the text. In most cases the complete details in determining the solutions are given while in the remainder of the problems helpful hints are provided. The problems chosen in each section represent, wherever possible, the variety of applications that are covered in the written material of the text, thereby providing the student a complete set of examples from which to learn.

Students should be aware that following these solutions is very different from designing and constructing one's own solution. Using this supplemental resource appropriately for learning differential equations is outlined as follows:

1. Make an honest attempt to solve the problem without using the guide.
2. If needed, glance at the beginning of the solution in the guide and then try again to generate the complete solution. Continue using the guide for hints when you reach an impasse.
3. Compare your final solution with the one provided to see whether yours is more or less efficient than the guide, since there is frequently more than one correct way to solve a problem.
4. Ask yourself why that particular problem was assigned.

In order to simplify the text, the following abbreviations have been used:

D.E.	differential equation(s)
O.D.E.	ordinary differential equation(s)
P.D.E.	partial differential equation(s)
I.C.	initial condition(s)
I.V.P.	initial value problem(s)
B.C.	boundary condition(s)
B.V.P.	boundary value problem(s)

I wish to express my appreciation to Dr. Joseph S. Torok for his

assistance in the preparation of the figures and to Mrs. Susan
A. Hickey for her excellent typing and proofreading of all
stages of the manuscript.

Charles W. Haines
Professor of Mathematics and
    Mechanical Engineering
Rochester Institute of Technology
Rochester, New York
November, 1991

CONTENTS

CHAPTER 1

Section 1.1, Page 9

2. The D.E. is second order since there is a second derivative of y appearing in the equation. The equation is nonlinear due to the y^2 term.

6. This is a third order D.E. since the highest derivative is y''' and it is linear since y and all its derivatives appear to the first power only. The terms x^3 and $\cos^2 x$ do not affect the linearity of the D.E.

8. For $y_1(x) = e^{-3x}$ we have $y_1'(x) = -3e^{-3x}$ and $y_1''(x) = 9e^{-3x}$. Substitution of these into the D.E. yields
$$9e^{-3x} + 2(-3e^{-3x}) - 3(e^{-3x}) = (9-6-3)e^{-3x} = 0.$$

14. Recall that if $u(x) = \int_0^x f(t)\,dt$, then $u'(x) = f(x)$.

16. Differentiating e^{rx} twice and substituting into the D.E. yields $r^2 e^{rx} - e^{rx} = (r^2-1)e^{rx}$. If $y = e^{rx}$ is to be a solution of the D.E. then the last quantity must be zero for all x. Thus $r^2-1 = 0$ since e^{rx} is never zero.

19. Differentiating x^r twice and substituting into the D.E. yields $x^2[r(r-1)x^{r-2}] + 4x[rx^{r-1}] + 2x^r = [r^2+3r+2]x^r$. If $y = x^r$ is to be a solution of the D.E., then the last term must be zero for all x and thus $r^2 + 3r + 2 = 0$.

24. The D.E. is second order since there are second partial derivatives of $u(x,y)$ appearing. The D.E. is nonlinear due to the product of $u(x,y)$ times u_x(or u_y).

28. Since $\dfrac{\partial u_1}{\partial t} = -\alpha^2 e^{-x^2 t}\sin x$ and $\dfrac{\partial^2 u_1}{\partial x^2} = -e^{-\alpha^2 t}\sin x$ we have $\alpha^2[-e^{-\alpha^2 t}\sin x] = -\alpha^2 e^{-\alpha^2 t}\sin x$, which is true for all t and x.

33. Observing the direction field,
     we see that for y>-1/2 we have
     y'<0, so the solution is
     decreasing here. Likewise,
     for y<-1/2 we have y'>0 and
     thus y(x) is increasing here.
     Since the slopes get closer as
     y gets closer to -1/2, we
     conclude that y→-1/2 as x→∞.

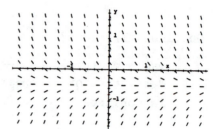

37.

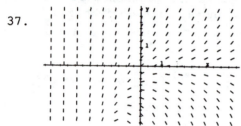

38.

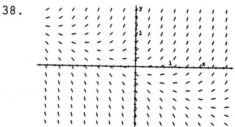

41. The isoclines are given by
     3-2y = c, or y = k, which are
     lines parallel to the x-axis.

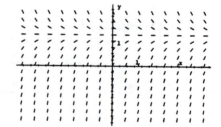

45. The isoclines are the circles
     x^2+y^2 = c, c > 0. This means
     that for c = 1 the solution
     will cross the circle x^2+y^2 = 1
     with a slope of 1, while for
     c = 2 the solution will cross
     the circle x^2+y^2 = 2 with a
     slope of 2.

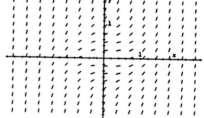

CHAPTER 2

<u>Section 2.1, Page 22</u>

1. $\mu(x) = \exp(\int 3dx) = e^{3x}$. Thus $e^{3x}(y'+3y) = e^{3x}(x+e^{-2x})$ or $\frac{d}{dx}(ye^{3x}) = xe^{3x} + e^x$. Integration of both sides yields

    $ye^{3x} = \frac{1}{3}xe^{3x} - \frac{1}{9}e^{3x} + e^x + c$, and division by e^{3x} gives

    the general solution. Note that $\int xe^{3x}dx$ is evaluated by integration by parts, with $u = x$ and $dv = e^{3x}dx$.

2. $\mu(x) = e^{-2x}$. 3. $\mu(x) = e^x$.

4. $\mu(x) = \exp(\int \frac{dx}{x}) = e^{\ln x} = x$, so $(xy)' = 3x\cos2x$, and integration by parts yields the general solution.

6. The equation must be divided by x so that it is in the form of Eq.(2): $y' + (2/x)y = (\sin x)/x$. Thus

    $\mu(x) = \exp(\int \frac{2dx}{x} = x^2$, and $(x^2y)' = x\sin x$. Integration then yields $x^2y = -x\cos x + \sin x + c$.

7. $\mu(x) = e^{x^2}$. 8. $\mu(x) = \exp(\int \frac{4xdx}{1+x^2}) = (1+x^2)^2$.

9. $\mu(x) = e^{-x}$ and $y = 2(x-1)e^{2x} + ce^x$. To find the value for c, set $x = 0$ in y and equate to 1, the initial value of y. Thus $-2+c = 1$ and $c = 3$, which yields the solution of the given initial value problem.

11. $\mu(x) = \exp(\int \frac{2dx}{x}) = x^2$ and $y = x^2/4 - x/3 + 1/2 + c/x^2$. Setting $x = 1$ and $y = 1/2$ we have $c = 1/12$.

14. $\mu(x) = x^2$. Thus $(x^2y)' = x\sin x$ and $x^2y = -x\cos x + \sin x + c$. Setting $x = \pi/2$ and $y = 1$ yields $c = \pi^2/4 - 1$.

16. $\mu(x) = xe^x$.

19.

21. The D.E. as given is nonlinear. However, if we think of
     y as the independent variable and x as the dependent
     variable then the D.E. can be written as $\dfrac{dx}{dy} = e^y - x$ or
     $\dfrac{dx}{dy} + x = e^y$. The integrating factor is then $\mu(y) = e^y$
     and the I.C. is $x(0) = 1$.

22a. To show that $\phi(x) = e^{2x}$ is a solution of the D.E., take
     its derivative and substitute into the D.E.

23. $[c\,\phi(x)]' + p(x)\,[c\phi(x)] = c[\phi'(x) + p(x)\phi(x)] = 0$ since
     $\phi(x)$ satisfies the given D.E.

24. $[y_1(x) + y_2(x)]' + p(x)[y_1(x) + y_2(x)] =$
     $y_1'(x) + p(x)y_1(x) + y_2'(x) + p(x)y_2(x) = 0 + g(x)$.

25. This problem demonstrates the central idea of the method
     of variation of parameters for the simplest case. The
     solution (ii) of the homogeneous D.E. is extended to the
     corresponding nonhomogeneous D.E. by replacing the
     constant A by a function A(x), as shown in (iii).

26. Assume $y(x) = A(x)\exp(-\!\int(-2)\,dx) = A(x)e^{2x}$.
     Differentiating y(x) and substituting into the D.E.
     yields $A'(x) = x^2$ since the terms involving A(x) add to
     zero. Thus $A(x) = x^3/3 + c$, which substituted into y(x)
     yields the solution.

27. $y(x) = A(x)\exp(-\!\int\dfrac{dx}{x}) = A(x)/x$.

Section 2.2, Page 28

Problems 1 through 4 follow the pattern of solution from
Section 2.1

2.   Remember to divide both sides of the D.E. by x^2 to get

$$\mu(x) = \exp\left(\int \frac{3dx}{x}\right) = e^{3\ln x} = x^3.$$

3. $\mu(x) = \exp\left(\int \tan x\, dx\right) = \exp\left(\int \frac{\sin x\, dx}{\cos x}\right)$

$\qquad = \exp[(-\ln(\cos x))] = \exp[\ln(1/\cos x)]$

$\qquad = \sec x.$

Multiplying both sides of the D.E. by secx and
simplifying yields $(y\sec x)' - x\sin x$. To solve this the
right side must be integrated by parts with u = x and
dv = sinxdx. Thus $y\sec x = -2x\cos x + 2\sin x + c$, which
yields the solution. For all steps we must have
$-\pi/2 < x < \pi/2$ for the functions to be defined.

In problems 5 through 12, we must determine the largest
interval in which the functions p and g are continuous and
which contain the initial point, and also determine the
constant c of the general solution. The procedure follows
Example 1 of this section.

5. Writing the D.E. in the form of Eq.(1) of this section we
have $y' + (2/x)y = x-1 + 1/x$, $y(1) = 1/2$. Thus p(x) and
g(x) are continuous on any interval not containing the
origin. Since the initial point is 1, the solution will
be valid on $0 < x < \infty$. $\mu(x) = x^2$ and thus
$(x^2 y)' = x^3 - x^2 + x$ and $y = \frac{1}{4}x^2 - \frac{1}{3}x + \frac{1}{2} + \frac{c}{x}.$
Substituting x = 1 and y = 1/2, we obtain c = 1/12, which
gives the desired solution.

7. p(x) = cotx and g(x) = 2cscx are both continuous for
$n\pi < x < (n+1)\pi$, where n is any integer. Since the
initial point is $x = \pi/2$, we choose n = 0 and conclude
that the solution will be valid on $0 < x < \pi$. Now
$\mu(x) = \exp\left(\int \cot x\, dx\right) = \sin x$ and thus $(y\sin x)' = 2$, which
gives the general solution $y = 2x\csc x + c\csc x$. Setting
$x = \pi/2$ and y = 1 we find $c = 1-\pi$.

10. $p(x) = \frac{2(1+x)}{x(2+x)}$ and $g(x) = \frac{1+3x}{x(2+x)}$. Thus p(x) and g(x)
are continuous on $-\infty < x < -2$, $-2 < x < 0$ and $0 < x < \infty$.
Since we are given y(-1) = 1, the solution will be valid
on $-2 < x < 0$. $\mu(x) = x^2 + 2x.$

11. p(x) and g(x) are continuous for all x and thus so is the
solution. $\mu(x) = e^x.$

12. $\mu(x) = (1 - x^2)^{1/2}$.

13. $\mu(x) = x^2$

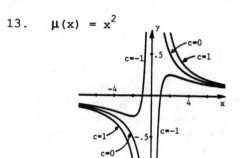

14. $\mu(x) = 1/x$

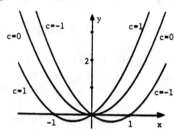

15. $\mu(x) = 1/x$. Solution
      exists only for $x > 0$
      since $g(x) = x^{1/2}$ and
      $p(x) = 1/x$ are defined
      and continuous only for
      $x > 0$.

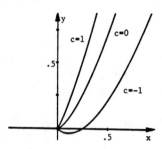

16. $\mu(x) = x$

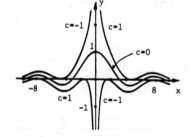

17. If the equation is written in the form of Eq.(1), then
    $p(x) = (\ln x)/(x-3)$ and $g(x) = 2x/(x-3)$. These are
    defined and continuous on the intervals $(0,3)$ and
    $(3,\infty)$, but since the initial point is $x = 1$, the
    solution will be continuous on $0 < x < 3$.

21. The general solution of the D.E. is $y = -2 + ce^x$.

23b. The limit as $x \rightarrow \infty$ of $[\dfrac{\sqrt{\pi}}{2} \text{erf}(x) + y(0)]$ must be zero

     for $\lim\limits_{x \rightarrow \infty} y$ to be finite. In this case L'Hopital's Rule
     applies.

25. $\mu(x) = e^{ax}$ so the D.E. can be written as
      $(e^{ax}y)' = be^{ax}e^{-\lambda x} = be^{(a-\lambda)x}$. If $a \neq \lambda$, then integration

and solution for y yields $y = [b/(a-\lambda)]e^{-\lambda x} + ce^{-ax}$. Then $\lim_{x \to \infty} y$ is zero since both λ and a are positive numbers.

If $a = \lambda$, then the D.E. becomes $(e^{ax}y)' = b$, which yields $y = (bx + c)/e^{\lambda x}$ as the solution. L'Hopital's Rule gives

$$\lim_{x \to \infty} y = \lim_{x \to \infty} \frac{(bx+c)}{e^{\lambda x}} = \lim_{x \to \infty} \frac{b}{\lambda e^{\lambda x}} = 0.$$

26a. $\mu(x) = e^{2x}$. Since g(x) is continuous on the interval $0 \le x \le 1$ we may solve the I.V.P.

$y_1' + 2y_1 = 1$, $y_1(0) = 0$ on that interval to obtain

$y_1 = 1/2 - (1/2)e^{-2x}$, $0 \le x \le 1$. g(x) is also continuous

for $1 < x$; and hence we may solve $y_2' + 2y_2 = 0$ to obtain

$y_2 = ce^{-2x}$, $1 < x$. The solution y_1 of the original
I.V.P. must be continuous (since its derivative must
exist) and hence we must choose c in y_2 so that y_2 at 1
has the same value as y_1 at 1. Thus

$ce^{-2} = 1/2 - e^{-2}/2$ or $c = (1/2)(e^2-1)$ and we obtain

$$y = \begin{cases} 1/2 - (1/2)e^{-2x} & 0 \le x \le 1 \\ 1/2(e^2-1)e^{-2x} & 1 \le x \end{cases} \qquad \text{and}$$

$$y' = \begin{cases} e^{-2x} & 0 \le x \le 1 \\ (1-e^2)e^{-2x} & 1 < x. \end{cases}$$

Evaluating the two parts of y' at $x_0 = 1$ we see that they
are different, and hence y' is not continuous at $x_0 = 1$.

27a. For $n = 0,1$, the D.E. is linear and Eqs.(3) and (4)
apply.

27b. Let $v = y^{1-n}$ then $\dfrac{dv}{dx} = (1-n)y^{-n}\dfrac{dy}{dx}$ so $\dfrac{dy}{dx} = \dfrac{1}{1-n}y^n\dfrac{dv}{dx}$,
which makes sense when $n \ne 0,1$. Substituting into the
D.E. yields $\dfrac{y^n}{1-n}\dfrac{dv}{dx} + p(x)y = q(x)y^n$ or
$v' + (1-n)p(x)y^{1-n} = (1-n)q(x)$. Setting $v = y^{1-n}$ then
yields a linear D.E. for v.

28. $n = 3$ so $v = y^{-2}$ and $\dfrac{dv}{dx} = -2y^{-3}\dfrac{dy}{dx}$ or $\dfrac{dy}{dx} = -\dfrac{1}{2}\,y^3\dfrac{dv}{dx}$.

Substituting this into the D.E. gives

$-\dfrac{1}{2}y^3\dfrac{dv}{dx} + \dfrac{2}{x}y = \dfrac{1}{x^2}y^3$. Simplifying and setting

$y^{-2} = v$ then gives the linear D.E.

$v' - \dfrac{4}{x}v = -\dfrac{2}{x^2}$, where $\mu(x) = \dfrac{1}{x^4}$ and

$v(x) = cx^4 + \dfrac{2}{5x} = \dfrac{2+5cx^2}{5x}$. Thus $y = \pm[5x/(2+5cx^2)]^{1/2}$.

29. $n = 2$ so $v = y^{-1}$ and $\dfrac{dv}{dx} = -y^2\dfrac{dv}{dx}$. Thus the D.E.

becomes $-y^2\dfrac{dv}{dx} - ry = -ky^2$ or $\dfrac{dv}{dx} + rv = k$. Hence

$\mu(x) = e^{rx}$ and $v = k/r + ce^{-rx}$. $y = 1/v$ then yields the solution.

Section 2.3, Page 35

Problems 1 through 16 follow the pattern of the examples worked in this section. The first eight problems, however, do not have I.C. so the integration constant, c, cannot be found.

1. Write the equation in the form $ydy = x^2dx$. Integrating the left side with respect to y and the right side with respect to x yields

$\dfrac{y^2}{2} = \dfrac{x^3}{3} + C$, or $3y^2 - 2x^3 = c$.

4. Factor the right side to obtain $y' = (1+x)(1+y^2)$, which separates to $((1+y^2)^{-1}dy = (1+x)dx$. Integrating each

side, as in problem 1, yields $\arctan y = \dfrac{x^2}{2} + x + c$,

which can be solved explicitly for y.

6. Separating the variables we get $(1-y^2)^{-1/2}dy = x^{-1}dx$.
Integrating each side yields $\arcsin y = \ln|x|+c$, so
$y = \sin[\ln|x|+c]$, $x \neq 0$.

11. Separate variables by factoring the denominator of the

right side to get $ydy = \dfrac{2x}{1+x^2}dx$.   Integration yields

$y^2/2 = \ln(1+x^2)+c$ and use of the I.C. gives $c = 2$.   Thus
$y = \pm [2\ln(1+x^2)+4]^{1/2}$, but we must discard the plus
square root because of the I.C.   Since $1 + x^2 > 0$, the
solution is valid for all x.

13.   Separating variables and integrating yields
$y + y^2 = x^2 + c$. Setting $y = 0$ when $x = 2$ yields $c = -4$
or $y^2 + y = x^2-4$.   To solve for y complete the square on
the left side by adding $1/4$ to both sides. This yields
$y^2 + y + \dfrac{1}{4} = x^2 - 4 + \dfrac{1}{4}$ or $(y + \dfrac{1}{2})^2 = x^2 - 15/4$.   Taking
the square root of both sides yields
$y + \dfrac{1}{2} = \pm\sqrt{x^2 - 15/4}$, where the positive square root
must be taken in order to satisfy the I.C.   Thus
$y = -\dfrac{1}{2} + \sqrt{x^2 - 15/4}$, which is defined for $x^2 \geq 15/4$ or
$x \geq \sqrt{15/2}$. The possibility that $x < -\sqrt{15/2}$ is
discarded due to the I.C.

15.   As above we start with $\cos 3ydy = -\sin 2xdx$ and integrate
to get $\dfrac{1}{3}\sin 3y = \dfrac{1}{2}\cos 2x + c$.   Setting $y = \pi/3$ when

$x = \pi/2$ (from the I.C.) we find that $0 = -\dfrac{1}{2} + c$ or

$c = \dfrac{1}{2}$, so that $\dfrac{1}{3}\sin 3y = \dfrac{1}{2}\cos 2x + \dfrac{1}{2} = \cos^2 x$ (using the
appropriate trigonometric identity).   To solve for y we
must choose the branch that passes through the point
$(\pi/2, \pi/3)$ and thus $3y = \pi - \arcsin(3\cos^2 x)$, or
$y = \dfrac{\pi}{3} - \dfrac{1}{3}\arcsin(3\cos^2 x)$, which is defined only for
$0 \leq 3\cos^2 x \leq 1$, or $-\sqrt{1/3} \leq \cos x \leq \sqrt{1/3}$.   Taking the
indicated square roots and then finding the inverse
cosine of each side yields $.9553 \leq x \leq 2.1863$, or
$|x-\pi/2| \leq 0.6155$, as the approximate interval.

17.   We have $(3y^2-6y)dy = (1+3x^2)dx$ so that $y^3-3y^2 = $
$x + x^3 - 2$, once the I.C. are used. $dx/dy = 0$ implies
$3y^2 - 6y = 0$, or $y = 0,2$.   For $y = 0$ we have
$x^3 + x - 2 = 0$, which is satisfied for $x = 1$, which is

the only zero of the function $w = x^3 + x - 2$. Likewise, for $y = 2$, $x = -1$.

20. Assume $cx + d \neq 0$ and $c \neq 0$, then use long division to obtain $\dfrac{dy}{dx} = \dfrac{a}{c} + (b - \dfrac{ad}{c})\dfrac{1}{cx+d} = \dfrac{a}{c} + (\dfrac{bc-ad}{c})\dfrac{1}{cx+d}$.
Integration then yields
$$y = \frac{a}{c}x + (\frac{bc-ad}{c^2})\ln|cx+d| + k.$$

22. If $v = y/x$ then $y = vx$ and $\dfrac{dy}{dx} = v + x\dfrac{dv}{dx}$ and thus the

D.E. becomes $v + x\dfrac{dv}{dx} = \dfrac{vx-4x}{x-vx} = \dfrac{v-4}{1-v}$. Subtracting v

from both sides yields $x\dfrac{dv}{dx} = \dfrac{v^2-4}{1-v}$, which separates into

$\dfrac{1-v}{v^2-4}dv = \dfrac{1}{x}dx$. To integrate the left side use partial

fractions to write $\dfrac{1-v}{v-4} = \dfrac{A}{v-2} + \dfrac{B}{v+2}$, which yields

$A = -1/4$ and $B = -3/4$. Integration then gives

$-\dfrac{1}{4}\ln|v-2| - \dfrac{3}{4}\ln|v+2| = \ln|x| - k$, or

$\ln|x^4||v-2||v+2|^3 = 4k$ after manipulations using
properties of the ln function. Setting $v = y/x$ and
further algebraic manipulations yield
$(y-2x)(y+2x)^3 = c$, where $c = e^{4k}$.

23a. The maximum and minimum values of $y/(1+2y^2)$ occur at
$y = +\sqrt{1/2}$ and $y = -\sqrt{1/2}$ respectively. Thus
$-\dfrac{1}{2\sqrt{2}} \leq \dfrac{y}{1+2y^2} \leq \dfrac{1}{2\sqrt{2}}$ and we know that $|\cos x| \leq 1$, hence
$f(x,y)$ is bounded as stated.

23b. Since $dy/dx = f(x,y)$, we have from part (a)
$-\dfrac{1}{2\sqrt{2}} \leq \dfrac{dy}{dx} \leq \dfrac{1}{2\sqrt{2}}$. Separating variables and using
the format of Eq.(15) of this section yields
$-\displaystyle\int_1^\phi \dfrac{dt}{2\sqrt{2}} \leq \int_1^\phi dy \leq \int_0^x \dfrac{dt}{2\sqrt{2}}$. Integration and evaluation

gives $\dfrac{-x}{2\sqrt{2}} \leq \phi(x) - 1 \leq \dfrac{x}{2\sqrt{2}}$ or $|\phi(x) - 1| \leq \dfrac{|x|}{2\sqrt{2}}$ for
all x. Geometrically, this says that the solution is

bounded by two straight lines of slope $\pm\dfrac{1}{2\sqrt{2}}$ passing

through the initial point $x = 0$, $y = 1$. Thus the
solution is bounded for all finite values of x.

Section 2.4, Page 41

2. Theorem 2.4.1 guarantees a unique solution to the D.E.
    through any point (x_0, y_0) such that $x_0^2 + y_0^2 < 1$ since
    $\dfrac{\partial f}{\partial y} = -y(1-x^2-y^2)^{1/2}$ is defined and continuous only for
    $1-x^2-y^2 > 0$. Note also that $f = (1-x^2-y^2)^{1/2}$ is defined
    and continuous in this region as well as on the boundary
    $x^2+y^2 = 1$. The boundary can't be included in the final
    region due to the discontinuity of $\dfrac{\partial f}{\partial y}$ there.

3. We are guaranteed a solution to the D.E. passing through
    any point in the xy plane since $f = 2xy/(1+y^2)$ and $\dfrac{\partial f}{\partial y} =$
    $2x(1-y^2)/(1+y^2)^2$ are defined and continuous for all x and
    y.

7. In this case $f = \dfrac{1+x^2}{y(3-y)}$ and $\dfrac{\partial f}{\partial y} = \dfrac{1+x^2}{y(3-y)^2} - \dfrac{1+x^2}{y^2(3-y)}$,
    which are both continuous everywhere except for $y = 0$ and
    $y = 3$.

9. The D.E. may be written as $y\,dy = -4x\,dx$ so that
    $\dfrac{y^2}{2} = -2x^2+c$, or $y^2 = c-4x^2$. The I.C. then yields
    $y_0^2 = c$, so that $y^2 = y_0^2 - 4x^2$ or $y = \pm\sqrt{y_0^2-4x^2}$, which is
    defined for $4x^2 < y_0^2$ or $|x| < |y_0|/2$. Note that $y_0 \neq 0$
    since Theorem 2.4.1 does not hold there.

14a. For $y_1 = 1-x$, $y_1' = -1 = \dfrac{-x+[x^2+4(1-x)]^{1/2}}{2}$

$$= \dfrac{-x+[(x-2)^2]^{1/2}}{2}$$

$$= \dfrac{-x+|x-2|}{2} = -1 \text{ if}$$

$(x-2) \geq 0$, by the definition of absolute value. Setting x = 2 in y_1 we get $y_1(2)$ = -1, as required.

14b. By Theorem 2.4.1 we are guaranteed a unique solution only
where $f(x,y) = \dfrac{-x+(x^2+4y)^{1/2}}{2}$ and $f_y(x,y) = (x^2+4y)^{-1/2}$ are
continuous. In this case the initial point $(2,-1)$ lies
in the region $x^2 + 4y \leq 0$, in which case $\dfrac{\partial f}{\partial y}$ is not
continuous and hence the theorem is not applicable and
there is no contradiction.

14c. If $y = y_2(x)$ then we must have $cx + c^2 = -x^2/4$, which is
not possible since c must be a constant.

Section 2.5, Page 49

1a. Comparing this problem with Eq.(1), we see that
r = .0525 and thus Eq.(8) yields τ = ln2/.0525 = 13.20
years as the half life of plutonium 241.

1b. Solving dQ/dt = -0.0525Q with Q(0) = 50 mg we find
$Q(t) = 50e^{-0.0525t}$. Thus $Q(10) = 50e^{-0.525} = 29.6$ mg.

4a. From Ex.(1) r = 0.02828 and thus dQ/dt = -0.01818Q + 1
gives the rate of change of thorium 234 in mg/day. The
solution of this equation with the initial condition of
Q(0) = 100 mg can be found using methods of Section 2.1
and is $Q(t) = (100 - \dfrac{1}{0.02828})e^{-0.02828t} + \dfrac{1}{0.02828}$, or
$Q(t) = 64.64e^{-0.02828t} + 35.36$ mg.

4c. Setting Q(t) = 35.86 mg in the answer to part(a) and
solving for t yields $t = \dfrac{-\ln(.5/64.64)}{0.02828}$ = 171.9 days.

7a. Set S_0 = 0 in Eq.(16) (or solve Eq.(15) with S(0) = 0).

7b. Set r = .075, t = 40 and S(t) = $1,000,000 in the answer
to (a) and then solve for k.

7c. Set k = $2,000, t = 40 and S(t) = $1,000,000 in the
answer to (a) and then solve numerically for r.

9.   The rate of accumulation due to interest is .1S and the
rate of decrease is k dollars per year and thus
dS/dt = .1S - k, S(0) = $8,000. Solving this for S(t)

yields $S(t) = 8000e^{.1t} - 10k(e^{.1t}-1)$ and substitution of $t = 3$ gives $k = \$3,086.64$ per year.

10. Since we are assuming continuity, either convert the monthly payment into an annual payment or convert the yearly interest rate into a monthly interest rate for 240 months. Then proceed as in Prob. 9.

13. Let $p(t)$ be the population at any time and let k be the constant of proportionality. Then the rate of change of $p(t)$ is given by $dp/dt = kp$, with $p(0) = 6 \times 10^8$ and $p(300) = 2.8 \times 10^9$. The solution of the D.E. is $p(t) = ce^{kt}$, where c is found by setting $t = 0$: $p(0) = c = 6 \times 10^8$. Thus $p(t) = (6 \times 10^8)e^{kt}$. To find k, set $t = 300$ and $p(300) = 2.8 \times 10^9$, yielding: $e^{300k} = (2.8/6)10$ or $k = .005135$. Hence $p(t) = (6 \times 10^8)e^{.005135t}$ gives the population at any time t. Setting $p(t) = 2.5 \times 10^{10}$ and solving for t yields the number of years after 1650 when the greatest population is reached.

14. From Eq.(22) we have $190° = 70° + (200°-70°)e^{-k}$ or $k = -\ln\dfrac{120}{130} = \ln(13/12)$. Hence if $\theta(T) = 150°$ we get: $150° = 70° + 130°e^{-\ln(13/12)T}$. Solving for T yields $T = \ln(8/13)/-\ln(13/12) = \ln(13/8)/\ln(13/12)$.

16. To determine the cooling rate, the information relevant to the morgue is used in Eq.(22): $\theta(t) = 40° + (85° - 40°)e^{-kt}$. Setting $t = 1$ and $Q(1) = 60$ and solving for k yields $k = .8109 (\text{hour})^{-1}$. To find the time of death, use Eq.(22) with $T = 70°$ and $\theta_0 = 98.6°$: $\theta(t) = 70° + 28.6e^{-.8109t}$. Setting $t = T$ and $\theta(T) = 85$ and solving for T gives $T = .7959$ hours before midnight, or 11:12 P.M.

17. The D.E. expressing the evaporation is $dV/dt = kS$, where the volume $V = \dfrac{4}{3}\pi r^3$ and the surface area $S = 4\pi r^2$. The D.E., in terms of r, then is $r^2\dfrac{dr}{dt} = kr^2$, with $r(0) = 3$ and $r(1) = 2$. When $r \neq 0$ the D.E. becomes $r' = k$, or $r(t) = kt + c$. Recall that r must be positive for the problem to have physical meaning.

18. Let $S(t)$ be the amount of salt that is present at any time t, then $S(0) = 0$ is the original amount of salt in the tank, 2γ is the amount of salt entering per minute, and $2(S/120)$ is the amount of salt leaving per minute (all amounts measured in grams). Thus
$dS/dt = 2\gamma - 2S/120$, $S(0) = 0$.

20. In order to find the amount of salt at the end of 20 minutes, we must first find the amount of salt that is present after 10 minutes. For the first 10 minutes (if we let $Q(t)$ be the amount of salt in the tank):
$$\frac{dQ}{dt} = \frac{1}{2}(2) - 2\frac{Q(t)}{100}, \quad Q(0) = 0.$$ This I.V.P. has the solution: $Q(t) = 50(1-e^{-.02t})$, which yields $Q(10) = 9.063$ lbs. of salt in the tank after the first 10 minutes. At this point no more salt is allowed to enter, so the new I.V.P. (letting $P(t)$ be the amount of salt in the tank) is: $\frac{dP}{dt} = (0)(2) - 2\frac{P(t)}{100}$,
$P(0) = Q(10) = 9.063$. The solution of this problem is $P(t) = 9.063e^{-.02t}$, which yields $P(10) = 7.42$ lbs. present 20 minutes after the start.

21. Salt flows into the tank at the rate of $(1)(3)$ lb/min. and it flows out of the tank at the rate of $\frac{Q(t)}{200+t}(2)$ lb/min. since the volume of water in the tank at any time t is $200 + (1)(t)$ gallons (due to the fact that water flows into the tank faster than it flows out). Thus the I.V.P. is $dQ/dt = 3 - \frac{2}{200+t}Q(t)$, $Q(0) = 100$.

22. Hint: let $Q(t)$ be the quantity of carbon monoxide in the room at any time t. Then the concentration is given by $x(t) = Q(t)/1200$.

23a. The required I.V.P. is $dQ/dt = kr + P - \frac{Q(t)}{V}r$,
$Q(0) = Vc_0$. Since $c = Q(t)/V$, the I.V.P. may be rewritten $Vc'(t) = kr + P - rc$, $c(0) = c_0$, which has the solution $c(t) = k + \frac{P}{r} + (c_0 - k - \frac{P}{r})e^{-rt/V}$.

23b. Set $k = 0$, $P = 0$, $t = T$ and $c(T) = .5c_0$ in the solution found in (a).

Problems 1 through 13 follow the pattern illustrated in Fig.2.6.2 and the discussion following Eq.(9).

3. The critical points are found by setting $\frac{dN}{dt}$ equal to zero. Thus $N = 0,1,2$ are the critical points. The graph of $N(N-1)(N-2)$ is positive for $0 < N < 1$ and $2 < N$ and negative for $1 < N < 2$. Thus $N(t)$ is increasing $(\frac{dN}{dt} > 0)$ for $0 < N < 1$ and $2 < N$ and decreasing $(\frac{dN}{dt} < 0)$ for $1 < N < 2$. Therefore 0 and 2 are unstable critical points while 1 is a stable critical point.

6. $\frac{dN}{dt}$ is zero only when $\text{Arctan} N$ is zero. $\frac{dN}{dt} > 0$ for $N < 0$ and $\frac{dN}{dt} < 0$ for $N > 0$. Thus $N = 0$ is a stable critical point.

7c. Separate variables to get $\frac{dN}{(1-N)^2} = kt$. Integration yields $\frac{1}{1-N} = kt + c$, or $N = 1 - \frac{1}{kt + c} = \frac{kt + c - 1}{kt + c}$. Setting $t = 0$ and $N(0) = N_0$ yields $N_0 = \frac{c-1}{c}$ or $c = \frac{1}{1-N_0}$. Hence $N(t) = \frac{(1-N_0)kt + N_0}{(1-N_0)kt + 1}$. For $N_0 > 1$ notice that the denominator will have a zero for some value of t, depending on the values chosen for N_0 and k. Thus the solution has a discontinuity at that point.

9. Setting $\frac{dN}{dt} = 0$ we find $N = 0, \pm 1$ are the critical points. Since $\frac{dN}{dt} > 0$ for $N < -1$ and $N > 1$ while $\frac{dN}{dt} < 0$ for $-1 < N < 1$ we may conclude that $N = -1$ is stable, $N = 0$ is semistable, and $N = 1$ is unstable.

11. $N = b^2/a^2$ and $N = 0$ are the only critical points. For

$0 < N < b^2/a^2$, $\frac{dN}{dt} < 0$ and thus $N = 0$ is a stable

equilibrium point. For $N > b^2/a^2$, $dN/dt > 0$ and thus
$N = b^2/a^2$ is unstable.

14. If $F'(N_1) < 0$ then the slope of N_1 is negative at N_1 and
     thus $F(N) > 0$ for $N < N_1$ and $F(N) < 0$ for $N > N_1$ since
     $F(N_1) = 0$. Hence N_1 is a stable critical point. A
     similar argument holds for $F'(N_1) > 0$.

16b. The graph of $\frac{dN}{dt}$ vs N has a maximum point at $N = K/e$.

     Thus $\frac{dN}{dt}$ is positive and increasing for $0 < N < K/e$ and

     thus $N(t)$ is concave up for that interval. Similarly $\frac{dN}{dt}$
     is positive and decreasing for $K/e < N < K$ and thus $N(t)$
     is concave down for that interval.

16c. $\ln(K/N)$ is very large for small values of N and thus
     $(rN)\ln(K/N) > rN(1 - N/K)$ for small N. Since $\ln(K/N)$
     and $(1 - N/K)$ are both strictly decreasing functions of
     N and since $\ln(K/N) = (1 - N/K)$ only for $N = K$, we may
     conclude that $\frac{dN}{dt} = (rN)\ln(K/N)$ is never less than

     $\frac{dN}{dt} = rN(1 - N/K)$.

17a. If $u = \ln(N/K)$ then $N = Ke^u$ and $\frac{dN}{dt} = Ke^u\frac{du}{dt}$ so that the
     D.E. becomes $du/dt = -ru$.

18b. Use the results of Problem 14.

18d. Differentiate Y with respect to E.

19a. Set $\frac{dN}{dt} = 0$ and solve for N using the quadratic formula.

19b. Use the results of Problem 14.

19d. If $h > rk/4$ there are no critical points (see part a) and
     $\frac{dN}{dt} < 0$ for all t.

22a. If $z = x/n$ then $dz/dt = \dfrac{1}{n}\dfrac{dx}{dt} - \dfrac{x}{n^2}\dfrac{dn}{dt}$. Use the

Equations (i) and (ii) then give the I.V.P. (iii).

22b. Separate variables to get $\dfrac{dz}{z(1-z)} = -\beta dt$. Using partial

fractions this becomes $\dfrac{dz}{z} + \dfrac{dz}{1-z} = -\beta dt$. Integration

and solving for z yields the answer.

24a. Plot dx/dt vs x and observe that $x = p$ and $x = q$ are
critical points. Also note that $dx/dt > 0$ for
$x < \min(p,q)$ and $x > \max(p,q)$ while $dx/dt < 0$ for x
between $\min(p,q)$ and $\max(p,q)$. Thus $x = \min(p,q)$ is a
stable point while $x = \max(p,q)$ is unstable. To solve
the D.E., separate variables and use partial fractions to
obtain $\dfrac{1}{q-p}[\dfrac{dx}{q-x} - \dfrac{dx}{p-x}] = \alpha dt$. Integration and solving
for x yields the solution.

24b. $x = p$ is a semistable critical point and since $\dfrac{dx}{dt} > 0$,

$x(t)$ is an increasing function. Thus for $x(0) = 0$, $x(t)$
approaches p as $t \to \infty$. To solve the D.E., separate
variables and integrate.

Section 2.7, Page 72

1a. If $x(t)$ is the height above the ground, then the I.V.P.
for $v(t)$ is $dv/dt = -g$, $v(0) = 20$. Thus
$\dfrac{dx}{dt} = v(t) = 20 - gt$ and $x(t) = 20t - (g/2)t^2 + c$. Since
$x(0) = 30$, $c = 30$ and $x(t) = 20t - (g/2)t^2 + 30$. At the
maximum height $v(t_m) = 0$ and thus
$t_m = 20/9.8 = 2.04$ sec., which when substituted in the
equation for $x(t)$ yields the maximum height.

1b. At the ground $x(t_g) = 0$ and thus $20t_g - 4.9t_g^2 + 30 = 0$.

2. The I.V.P. in this case is $m\dfrac{dv}{dt} = -\dfrac{1}{30}v - mg$, $v(0) = 20$,
where the positive direction is measured upward.

5. The I.V.P. is $m(dv/dt) = mg - kv$, $v(0) = 0$ which has the

solution $v(t) = \dfrac{mg}{k}(1 - e^{-kt/m})$.   Setting

$v(t) = .9\dfrac{mg}{k}$ and solving for t yields the answer.

7a. The I.V.P. is $m\dfrac{dv}{dt} = mg - .75v$, $v(0) = 0$ and v is measured positively downward. The solution to this equation is $v(t) = 240(1-e^{-.133t})$ so that $v(10) = 176.7$ ft/sec.

7b. Integration of v(t) as found in (a) yields $x(t) = 240t + 1800(e^{-.133t}-1)$ where x is measured positively down from the altitude of 5000 feet. Set $t = 10$ to find the distance traveled when the parachute opens.

7c. After the parachute opens the I.V.P. is $m\dfrac{dv}{dt} = mg-12v$, $v(0) = 176.7$, which has the solution $v(t) = 161.7e^{-2.133t} + 15$ and where $t = 0$ now represents the time the parachute opens. Letting $t \to \infty$ yields the limiting velocity.

7d. Integrate v(t) as found in (c) to find $x(t) = 15t - 75.8e^{-2.133t} + C_2$. $C_2 = 75/8$ since $x(0) = 0$, x now being measured from the point where the parachute opens. Setting $x = 3925.5$ will then yield the length of time the skydiver is in the air after the parachute opens.

8. The I.V.P. is $m(dv/dt) = mg - k\sqrt{v}$; $v(0) = v_0$ where the positive direction is down. To solve, separate variables to obtain $\dfrac{mdv}{mg - k\sqrt{v}} = dt$ and let $u = mg - k\sqrt{v}$ to integrate the left side.

9. The I.V.P. is $m(dv/dt) = mg - kv^2$, $v(0) = 0$. To solve, separate variables and use partial fractions.

12b. From part (a) $v(t) = -\dfrac{mg}{k} + [v_0 + \dfrac{mg}{k}]e^{-kt/m}$. As $k \to 0$ this has the indeterminant form of $-\infty + \infty$. Thus rewrite $v(t)$ as $v(t) = [-mg + (v_0 k + mg)e^{-kt/m}]/k$ which has the

indeterminant form of 0/0, as k → 0 and hence
L'Hopital's Rule may be applied with k as the variable.

13a. The equation of motion is m(dv/dt) = w-R-B which, in this
problem, is

$$\frac{4}{3}\pi a^3\rho(dv/dt) = \frac{4}{3}\pi a^3\rho g - 6\pi\mu av - \frac{4}{3}\pi a^3\rho'g.$$ The limiting

velocity occurs when dv/dt = 0.

13b. Since the droplet is motionless, v = dv/dt = 0, we have

the equation of motion $0 = (\frac{4}{3})\pi a^3\rho g - Ee - (\frac{4}{3})\pi a^3\rho'g$,

where ρ is the density of the oil and ρ' is the density
of air. Solving for e yields the answer.

14. All three parts can be answered from one solution if k
represents the resistance and if the method of solution
of Example 2 is used. Thus we have

$$m\frac{dv}{dt} = mv\frac{dv}{dx} = mg - kv, \quad v(0) = 0,$$ where we have assumed

the velocity is a function of x. The solution of this
I.V.P. involves a logarithmic term, and thus the answers
to parts (a) and (c) must be found using a numerical
procedure.

15b. Note that 32 ft/sec^2 = 78,545 m/hr^2.

16. This problem is the same as Example 2 through Eq.(15).

In this case the I.C. is $v(\xi R) = v_o$, so $c = v_o^2 - \frac{2gR}{1+\xi}$.

v_e is found by noting that $v_o^2 \geq \frac{2gR}{1+\xi}$ in order for v^2 to

always be positive. From Example 2, the escape velocity

for a surface launch is $v_e(0) = \sqrt{2gR}$. We want the

escape velocity of x_o = R to have the relation

$v_e(\xi R) = .85v_e(0)$, which yields $\xi = (0.85)^{-2} - 1 \cong 0.384$.
If R = 4000 miles then $x_o = \xi R$ = 1536 miles.

Section 2.8, Page 80

3.   M(x,y) = $3x^2$-2xy+2 and N(x,y) = $6y^2$-x^2+3, so M_y= -2x = N_x
and thus the D.E. is exact. Integrating M(x,y) with
respect to x we get $\psi(x,y) = x^3 - x^2 + 2x + H(y)$. Taking
the partial derivative of this with respect to y and

setting it equal to N(x,y) yields $-x^2+h'(y) = 6y^2-x^2+3$, so
that $h'(y) = 6y^2 + 3$ and $h(y) = 2y^3 + 3y$. Substitute this
h(y) into $\psi(x,y)$ and recall that the equation which
defines y(x) implicitly is $\psi(x,y) = c$. Thus
$x^3 - x^2y + 2x + 2y^3 + 3y = c$ is the equation that yields
the solution.

5. Writing the equation in the form $M(x,y)dx + N(x,y)dy = 0$
gives $M(x,y) = ax + by$ and $N(x,y) = bx + cy$. Now
$M_y = b = N_x$ and the equation is exact. Integrating

M(x,y) with respect to x yields $\psi(x,y) = (a/2)x^2 + bxy + h(y)$. Differentiating ψ with respect to y (x constant)
and setting $\psi_y(x,y) = N(x,y)$ we find that $h'(y) = cy$ and

thus $h(y) = (c/2)y^2$. Hence the solution is given by
$(a/2)x^2 + bxy + (c/2)y^2 = k$.

6. The D.E. must be put into the form of Eq.(6).

7. $M_y(x,y) = e^x\cos y - 2\sin x = N_x(x,y)$ and thus the D.E. is
exact. Integrating M(x,y) with respect to x gives
$\psi(x,y) = e^x\sin y + 2y\cos x + h(y)$. Finding $\psi_y(x,y)$ from
this and setting that equal to N(x,y) yields $h'(y) = 0$
and thus h(y) is a constant. Hence an implicit solution
of the D.E. is $e^x\sin y + 2y\cos x = c$. The solution $y = 0$
is also valid since it satisfies the D.E. for all x.

9. If you try to find $\psi(x,y)$ by integrating M(x,y) with
respect to x you must integrate by parts. Instead find
$\psi(x,y)$ by integrating N(x,y) with respect to y to obtain
$\psi(x,y) = e^{xy}\cos 2x - 3y + g(x)$. Now find g(x) by
differentiating $\psi(x,y)$ with respect to x and set that
equal to M(x,y), which yields $g'(x) = 2x$ or $g(x) = x^2$.

12. As long as $x^2 + y^2 \neq 0$, we can simplify the equation by
multiplying both sides by $(x^2 + y^2)^{3/2}$. This gives the
exact equation $xdx + ydy = 0$. The solution to this
equation is given implicitly by $x^2 + y^2 = c$. If you
apply Theorem 2.8.1 and its construction without the
simplification, you get $(x^2 + y^2)^{-1/2} = C$ which can be
written as $x^2 + y^2 = c$ under the same assumption required
for the simplification.

14. $M_y = 1$ and $N_x = 1$, so the D.E. is exact. Integrating

M(x,y) with respect to x yields

$\psi(x,y) = 3x^3 + xy - x + h(y)$. Differentiating this with respect to y and setting $\psi_y(x,y) = N(x,y)$ yields

$h'(y) = -4y$ or $h(y) = -2y^2$. Thus the implicity solution is $3x^3 + xy - x - 2y^2 = c$. Setting $x = 1$ and $y = 0$ gives $c = 2$ so that $2y^2 - xy + (2+x-3x^3) = 0$ is the implicit solution satisfying the given I.C. Use the quadratic formula to find y(x), where the negative square root is used in order to satisfy the I.C.

15. We want $M_y(x,y) = 2xy + bx^2$ to be equal to

$N_x(x,y) = 3x^2 + 2xy$. thus we must have $b = 3$. This

gives $\psi(x,y) = \frac{1}{2}x^2y^2 + x^3y + h(y)$ and consequently

$h'(y) = 0$. After multiplying through by 2, the solution is given implicitly by $x^2y^2 + 2x^3y = c$.

19. $M_y(x,y) = 3x^2y^2$ and $N_x(x,y) = 1 + y^2$ so the equation is no exact by Theorem 2.8.1. Multiplying by the integrating factor $\mu(x,y) = 1/xy^3$ we get

$x + \frac{(1+y^2)}{y^3}y' = 0$, which is an exact equation since

$M_y = N_x = 0$ (it is also separable). In this case

$\psi = \frac{1}{2}x^2 + h(y)$ and $h'(y) = y^{-3} + y^{-1}$ so that

$x^2 - y^{-2} + 2\ln|y| = c$ gives the solution implicitly.

22. Multiplication of the given D.E. (which is not exact) by

$\mu(x,y) = xe^x$ yields $(x^2 + 2x)e^x\sin y\, dx + x^2e^x\cos y\, dy$, which is exact since $M_y(x,y) = N_x(x,y) = (x^2+2x)e^x\cos y$. To solve this exact equation it's easiest to integrate $N(x,y) = x^2e^x\cos y$ with respect to y to yield $\psi(x,y) = x^2e^x\sin y + g(x)$. Solving for g(x) yields the implicit solution.

23. This problem is similar to the derivation leading up to Eq.(26). Assuming that μ depends only on y, we find from Eq.(25) that $\mu' = Q\mu$, where $Q = (N_x - M_y)/M$ must depend on y alone. Solving this last D.E. yields $\mu(y)$ as given. This method provides an alternative approach to Problems 27 through 30.

25. The equation is not exact so we must attempt to find an integrating factor. Since $\dfrac{1}{N}(M_y-N_x) = \dfrac{3x^2 + 2x + 3y^2 - 2x}{x^2 + y^2} = 3$ is a function of x alone there is an integrating factor depending only on x, as shown in Eq.(26). Then $d\mu/dx = 3\mu$, and the integrating factor is $\mu(x) = e^{3x}$. Hence the equation can be solved as in Example 4.

26. An integrating factor can be found which is a function of x only, yielding $\mu(x) = e^{-x}$. Alternatively, you might recognize that $y' - y = e^{2x} - 1$ is a linear first order equation which can be solved as in Section 2.1.

27. Using the results of Problem 23, it can be shown that $\mu(y) = y$ is an integrating factor. Thus multiplying the D.E. by y gives $ydx + (x - y\sin y)dy = 0$, which can be identified as an exact equation. Alternatively, one can rewrite the last equation as $(ydx + xdy) - y\sin y\, dy = 0$. The first term is $d(xy)$ and the last can be integrated by parts. Thus we have $sy + y\cos y - \sin y = c$.

29. By multiplying by siny we obtain $e^x\sin y\, dx + e^x\cos y\, dy + 2y\, dy = 0$, and the first two terms are just $d(e^x\sin y)$. Thus $e^x\sin y + y^2 = c$.

31. Using the results of Problem 24, it can be shown that $\mu(xy) = xy$ is an integrating factor. Thus multiplying by xy we have $(3x^2y + 6x)dx + (x^3 + 3y^2)dy = 0$, which can be identifies as an exact equation. Alternatively, we can observe that the above equation can be written as $d(x^3y) + d(3x^2) + d(y^3) = 0$, so that $x^3y + 3x^2 + y^3 = c$.

Section 2.9, Page 84

Problems 1 through 8 are done in the same manner as the illustrative example of this section.

1. The D.E. can be written as $\dfrac{dy}{dx} = 1 + y/x$ and is thus homogeneous. Setting $v = y/x$ yields $x\dfrac{dv}{dx} + v = 1 + v$ so that $\dfrac{dv}{dx} = \dfrac{1}{x}$. This equation is separable and has the solution $\ln|x| = v + c$. Since $v = y/x$, we obtain $x\ln|x| = y + cx$ as the solution.

3. Writing the equation so that the right side is a function
    of y/x gives $dy/dx = 1 + (y/x) + (y/x)^2$ so the equation
    is homogeneous. The substitution y = vx leads to
    $v + x\dfrac{dv}{dx} = 1 + v + v^2$ or $\dfrac{dv}{1 + v^2} = \dfrac{dx}{x}$. Solving, we get
    arctanv = ln|x| + c. Substituting for v we obtain
    arctan(y/x) - ln|x| = c.

5. Dividing the numerator and denominator of the right side
    by x and substituting y = vx we get $v + x\dfrac{dv}{dx} = \dfrac{4v - 3}{2-v}$
    which can be rewritten as $x\dfrac{dv}{dx} = \dfrac{v^2 + 2v - 3}{2 - v}$. Separating
    variables gives $\dfrac{2 - v}{(v+3)(v-1)}\,dv = \dfrac{1}{x}dx$. Applying a
    partial fraction decomposition to the left side we obtain
    $[\dfrac{1}{4}\dfrac{1}{v-1} - \dfrac{5}{4}\dfrac{1}{v+3}]dv = \dfrac{dx}{x}$, and upon integrating both sides
    we find that $\dfrac{1}{4}\ln|v-1| - \dfrac{5}{4}\ln|v+3| = \ln|x| + c$.
    Substituting for v and performing some algebraic
    manipulations we get the solution in the implicit form
    $|y-x| = c|y+3x|^5$.

9b. Making the suggested substitution we obtain
    $\dfrac{dY}{dX} = \dfrac{2(Y-k) - (X-h) + 5}{2(X-h) - (Y-k) - 4} = \dfrac{2Y-X+(h-2k+5)}{2X-Y+(k-2h-4)}$, which will be
    homogeneous in X,Y provided the two expressions in
    parenthesis are both zero. Solving these yields h = -1
    and k = 2. From part(a), it then follows that
    $|Y-X| = c|Y+X|^3$. Substituting Y = y + 2, X = x - 1 gives
    the solution in the form $|y - x + 3| = c|y + x + 1|^3$.

10. The transformation y = Y - 1, x = X - 3 reduces the
    equation to Problem 6.

12. As in previous problems we can write the D.E. as
    $x\dfrac{dv}{dx} = -\dfrac{2v(v+2)}{v+1}$. Separating variables, using a partial
    fraction decomposition, and solving gives
    $|v|^{1/2}|v+2|^{1/2}x^2 = c$. Substituting for v and simplifying
    yields $x^2y^2 + 2x^3y = c$ which is the same as the answer
    found in Example 2, Section 2.9.

13. To show that $\mu(x,y)$ is an integrating factor, multiply both terms of the D.E. by the given $\mu(x,y)$. Since the D.E. is homogeneous, we know that $M(x,y)/N(x,y) = F(y/x)$ and thus the D.E. may be written as

$$\frac{F}{xF + y}dx + \frac{1}{xF + y}dy = 0 \text{ after multiplying both}$$

numerators and denominators by $1/N$ and substituting F for M/N. It may now be shown that $\dfrac{\partial}{\partial x}[\dfrac{1}{xF + y}] = \dfrac{\partial}{\partial y}[\dfrac{F}{xF + y}]$ and hence the equation is exact.

14b. As in Problem 13 we have

$$\mu(x,y) = \frac{1}{x^3 + 3xy^2 - 2xy^2} = \frac{1}{x(x^2 + y^2)}. \text{ Multiplying the}$$

D.E. by this $\mu(x,y)$ yields

$$\frac{x^2 + 3y^2}{x(x^2 + y^2)}dx - \frac{2xy}{x(x^2 + y^2)}dy = 0, \text{ which can be shown to be}$$

exact. Thus there exists a $\psi(x,y)$ such that

$$\psi_y(x,y) = -\frac{2y}{x^2 + y^2}, \text{ so upon integrating with respect to y}$$

we obtain $\psi(x,y) = -\ln(x^2 + y^2) + g(x)$. We must also

have $\psi_x(x,y) = -\dfrac{2x}{x^2 + y^2} + g'(x) = \dfrac{x^2 + 3y^2}{x(x^2 + y^2)}$, so

$$g'(x) = \frac{x^2 + 3y^2 + 2x^2}{x(x^2 + y^2)} = \frac{3}{x}. \text{ Thus } g(x) = 3\ln|x| + \ln c,$$

and $\psi(x,y) = -\ln(x^2 + y^2) + 3\ln|x| + \ln c$, or $x^3 = c(x^2 + y^2)$.

Section 2.10, Page 86

In Problems 1 through 32 the D.E. is identified as to type.

1. Linear   2. Homogeneous

3. Exact   4. Linear equation in x

5. Exact   6. Linear

7. Linear equation in u   8. Linear

9. Exact   10. Integrating factor
                                  depends on x only

11. Exact   12. Linear

13. Homogeneous	14. Exact or homogeneous
15. Separable	16. Homogeneous
17. Linear	18. Linear or homogeneous
19. Integrating factor depends on x only	20. Separable
21. Homogeneous	22. Separable
23. Bernoulli equation	24. Separable
25. Exact	26. Integrating factor depends on x only
27. Integrating factor depends on x only	28. Exact
29. Homogeneous	30. Linear equation in x
31. Separable	32. Integrating factor depends on y only

34. Differentiating the equation with respect to x gives $p = \dfrac{dy}{dx} = \dfrac{1}{p}\dfrac{dp}{dx}$. Separating variables gives $dx = \dfrac{1}{p^2}dp$ which has the solution $x = -\dfrac{1}{p} + c$. Hence the solution may be represented parametrically by $y = \ln p$, $x = -\dfrac{1}{p} + c$. To solve the equation in another manner write it as $p = \dfrac{dy}{dx} = e^y$ which separates to $e^{-y}dy = dx$ and we get the solution $x + e^{-y} = c$.

36b. Comparing the given D.E. with the D.E. of Problem 35 we see that in this case $q_2(x) = -1/x$ and $q_3(x) = 1$. Hence, using the method suggested in Problem 35 we must solve $\dfrac{dv}{dx} = -(-1/x + 2y_1)v - 1$, where $y_1(x)$ is given as $1/x$. This is a linear first-order equation which has the solution $v(x) = (\dfrac{c - x^2}{2})/x$ and thus a general solution to the Riccati equation is $y_2(x) = 1/x + 2x/(c - x^2)$.

39b. Following the procedure outlined in Problem 38, we first
find that the given family of curves has the slope

$$dy/dx = \frac{-x + c}{y}$$ at each point. To eliminate c, solve the

given equation for c to get $c = \frac{x^2 + y^2}{2x}$. Thus

$$dy/dx = \frac{y^2 - x^2}{2xy}$$ and the orthogonal family of curves will

then have slopes given by $dy/dx = \frac{-2xy}{y^2 - x^2}$. This equation

may be written as $y' = \frac{-2(y/x)}{(y/x)^2 - 1}$ so it is homogeneous.
Substitution of y = vx, separation of variables, and a
partial fraction decomposition lead to the equation
$(\frac{1}{v} - \frac{2v}{1+v^2})dv = \frac{dx}{x}$ whose solution is given implicitly by

$$\frac{v}{1 + v^2} = \hat{c}x.$$ Substitution for v and algebraic

simplification give $\frac{y}{x^2+y^2} = \hat{c}$ or $x^2 + y^2 = 2cy$.

Completion of the square gives the answer in more easily
recognizable form.

39d. The solution $y + \sqrt{x^2+y^2} = \tilde{c} x^2$ is obtained if the
procedure used in Problem 39b is followed. This answer
can be put in the desired (and preferable) form by
subtracting y from both sides and then squaring to obtain
$x^2+y^2 = (\tilde{c}x^2-y)^2 = \tilde{c}^2x^4 + 2\tilde{c}x^2y + y^2$. Subtracting y^2 and
division by x^2 (x ≠ 0) yields the answer.

40b. The slope of the given family of curves is given by
$m_2 = dy/dx = -x/y$ and $\tan\theta = 1$. Thus, from the given

relationship of θ, m_1 and m_2 we have $1 = \frac{m_1 + x/y}{1 - m_1(x/y)}$.

Solving for m_1, we find the intersecting family of
curves must have slopes $m_1 = dy/dx = (y-x)/(y+x)$, which
is a homogeneous D.E.

41. Using the equation for the slope of a line passing

through the points (x,y) and (a,b) we find the equation
to solve is y' = (y-b)/(x-a).

43. In this case the point through which the tangent line
must pass is (0,y/2) and the equation to solve is
y' = y/2x.

44. The D.E. is $S' = kS^2$, subject to the conditions $S(0) = 1$
million, $S(1) = 2$ million. Separating variables and
integrating leads to $S(t) = -\dfrac{1}{kt + c}$. The I.C. $S(0) = 1$
gives $c = -1$. Using the second condition, $S(1) = 2 =$
$-\dfrac{1}{k - 1}$ gives $k = \dfrac{1}{2}$ so the solution is $S(t) = \dfrac{2}{2 - t}$
million. In 6 months, $t = \dfrac{3}{2}$ (since $t = 0$ occurred 1
year ago) and $S(\dfrac{3}{2}) = 4$ million. As t approaches 2, his
wealth approaches infinity, so the solution cannot be
extended until two years from now (which would be $t = 3$).

45a. The D.E. is $dV/dt = k - \alpha\pi r^2$. The volume of a cone of
height L and radius r is given by $V = \pi r^2 L/3$ where
$L = hr/a$ from symmetry. Solving for r yields the desired
solution.

45b. From the material on logistic growth in Section 2.6 we
have equilibrium given by $k - \alpha\pi r^2 = 0$.

45c. $L \leq h$.

Section 2.11, Page 96

1. Let $s = x-1$ and $w(s) = y(x(s)) - 2$, then when $x = 1$ and
$y = 2$ we have $s = 0$ and $w(0) = 0$. Also,
$$\frac{dw}{ds} = \frac{dw}{dx} \cdot \frac{dx}{ds} = \frac{d}{dx}(y-2)\frac{dx}{ds} = \frac{dy}{dx} \text{ and hence}$$
$$\frac{dw}{ds} = (s+1)^2 + (w+2)^2, \text{ upon substitution into the given}$$
D.E.

5. Using Eq.(9) and following the steps of the Example of
this section we obtain the following:

$$\phi_1(x) = \int_0^x (t\phi_0(t) +1)\,dt = t\Big|_0^x = x$$

$$\phi_2(x) = \int_0^x (t^2+1)\,dt = (\frac{t^3}{3} + t)\Big|_0^x = x + \frac{x^3}{3}$$

$$\phi_3(x) = \int_0^x (t^2 + \frac{t^4}{3} + 1)\,dt = (\frac{t^3}{3} + \frac{t^5}{3\cdot 5} + t)\Big|_0^x = x + \frac{x^3}{3} + \frac{x^5}{3\cdot 5}.$$

Based upon these we hypothesize that:

$$\phi_n(x) = \sum_{k=1}^{n} \frac{x^{2k-1}}{1\cdot 3\cdot 5 \cdots (2k-1)} \quad \text{and use mathematical induction to}$$

verify this form for $\phi_n(x)$. Using Eq. (9) again we have:

$$\phi_{n+1}(x) = \int_0^x (\sum_{k=1}^{n} \frac{t^{2k}}{1\cdot 3\cdot 5 \cdots (2k-1)} + 1)\,dt$$

$$= \sum_{k=1}^{n} \frac{x^{2k+1}}{1\cdot 3\cdot 5 \cdots (2k+1)} + x$$

$$= \sum_{k=0}^{n} \frac{x^{2k+1}}{1\cdot 3\cdot 5 \cdots (2k+1)}$$

$$= \sum_{i=1}^{n+1} \frac{x^{2i-1}}{1\cdot 3\cdot 5 \cdots (2i-1)}, \quad \text{where } i = k+1. \text{ Since this is}$$

the same form for $\phi_{n+1}(x)$ as derived from $\phi_n(x)$ above, we
have verified by mathematical induction that $\phi_n(x)$ is as
given.

Section 2.12, Page 109

2. Using the given difference equation we have for n=0,
    $y_1 = y_0/2$; for n=1, $y_2 = 2y_1/3 = y_0/3$; and for n=2,
    $y_3 = 3y_2/4 = y_0/4$. Thus we guess that $y_n = y_0/(n+1)$, and
    the given equation then gives $y_{n+1} = \frac{n+1}{n+2} y_n = y_0/(n+2)$,
    which, by mathematical induction, verifies $y_n = y_0/(n+1)$
    as the solution for all n.

5. From the given equation we have $y_1 = .5y_0 + 6$.

    $$y_2 = .5y_1 + 6 = (.5)^2 y_0 + 6(1 + \frac{1}{2}) \text{ and}$$

    $$y_3 = .5y_2 + 6 = (.5)^3 y_0 + 6(1 + \frac{1}{2} + \frac{1}{4}). \text{ In general, then}$$

    $$y_n = (.5)^n y_0 + 6(1 + \frac{1}{2} + \cdots + \frac{1}{2^{n-1}})$$

    $$= (.5)^n y_0 + 6(\frac{1 - (1/2)^n}{1 - 1/2})$$

$$= (.5)^n y_0 + 12 - (.5)^n 12$$

$$= (.5)^n (y_0-12) + 12.$$ Mathematical induction can now be used to prove that this is the correct solution.

10. The governing equation is $y_{n+1} = \rho y_n - b$, which has the

solution $y_n = \rho^n y_0 - \dfrac{1-\rho^n}{1-\rho} b$ (Eq.(13) with a negative b).

Setting $y_{360} = 0$ and solving for b we obtain

$$b = \frac{(1-\rho)\rho^{360} y_0}{1-\rho^{360}}.$$

13. You must solve Eq.(13) numerically for ρ when $y_{240} = 0$, $b = -\$900$ and $y_0 = \$95,000$.

14. Substituting $u_n = \dfrac{\rho-1}{\rho} + v_n$ into Eq.(20) we get

$$\frac{\rho-1}{\rho} + v_{n+1} = \rho(\frac{\rho-1}{\rho} + v_n)(1 - \frac{\rho-1}{\rho} - v_n)\ \text{or}$$

$$v_{n+1} = -\frac{\rho-1}{\rho} + (\rho-1+\rho v_n)(\frac{1}{\rho} - v_n)$$

$$= \frac{1-\rho}{\rho} + \frac{\rho-1}{\rho} - (\rho-1)v_n + v_n - \rho v_n^2 = (2-\rho)v_n - \rho v_n^2.$$

15a. For $u_0 = .2$ we have $u_1 = 3.2u_0(1-u_0) = .512$ and
     $u_2 = 3.2u_1(1-u_1) = .7995392$. Likewise $u_3 = .51288406$,
     $u_4 = .7994688$, $u_5 = .51301899$, $u_6 = .7994576$ and
     $u_7 = .5130404$. Continuing in this fashion,
     $u_{14} = u_{16} = .79945549$ and $u_{15} = u_{17} = .51304451$.

17. For both parts of this problem a computer spreadsheet
     was used and an initial value of $u_0 = .2$ was chosen.
     Different initial values or different computer programs
     may need a slightly different number of iterations to
     reach the limiting value.

17a. The limiting value of .65517 (to 5 decimal places) is
     reached after approximately 100 iterations for $\rho = 2.9$.
     The limiting value of .66102 (to 5 decimal places) is
     reached after approximately 200 iterations for $\rho = 2.95$.
     The limiting value of .66555 (to 5 decimal places) is
     reached after approximately 910 iterations for $\rho = 2.99$.

17b. The solution oscillates between .63285 and .69938 after

approximately 400 iterations for ρ = 3.01. The solution
oscillates between .59016 and .73770 after approximately
130 iterations for ρ = 3.05. The solution oscillates
between .55801 and .76457 after approximately 30
iterations for ρ = 3.1. For each of these cases
additional iterations verified the oscillations were
correct to five decimal places.

18. For an initial value of .2 and ρ = 3.448 we have the
solution oscillating between .4403086 and .8497146.
After approximately 3570 iterations the eighth decimal
place is still not fixed, though. For the same initial
value and ρ = 3.45 the solution oscillates between the
four values: .43399155, .84746795, .44596778 and
.85242779 after 3700 iterations. For ρ = 3.449, the
solution is still varying in the fourth decimal place
after 3700 iterations, but there appear to be four
values.

CHAPTER 3

Section 3.1, Page 120

1. Assume $y = e^{rx}$, which is substituted into the D.E. to
    obtain the characteristic equation $r^2 + 2r - 3 = 0$.
    Hence $r_1 = 1$ and $r_2 = -3$ and the general solution is then
    $y = c_1 e^x + c_2 e^{-3x}$.

5. The characteristic equation is $r^2 + 5r = 0$, so the roots
    are $r_1 = 0$, and $r_2 = -5$. Thus
    $y = c_1 e^{ox} + c_2 e^{-5x} = c_1 + c_2 e^{-5x}$.

7. The characteristic equation is $r^2 - 9r + 9 = 0$ so that
    $r = (9 \pm \sqrt{81-36})/2 = (9 \pm 3\sqrt{5})/2$. Hence
    $y = c_1 \exp[(9+3\sqrt{5})x/2] + c_2 \exp[(9-3\sqrt{5})x/2]$.

10. Substituting $y = e^{rx}$ in the D.E.
    we obtain the characteristic
    equation $r^2 + 4r + 3 = 0$, which
    has the roots $r_1 = -1$, $r_2 = -3$.
    Thus $y = c_1 e^{-x} + c_2 e^{-3x}$ and
    $y' = -c_1 e^{-x} - 3c_2 e^{-3x}$.
    Substituting $x = 0$ we then have
    $c_1 + c_2 = 2$ and $-c_1 - 3c_2 = -1$,
    yielding $c_1 = 5/2$ and
    $c_2 = -1/2$. Thus $y = \dfrac{5}{2}e^{-x} - \dfrac{1}{2}e^{-3x}$
    and hence $y \to 0$ as $x \to \infty$.

13. The characteristic equation is $r^2 + 8r - 9 = 0$, so that
    $r_1 = 1$ and $r_2 = -9$ and the general solution is
    $y = c_1 e^x + c_2 e^{-9x}$. Since the I.C. are given at $x = 1$, it
    is convenient to introduce the arbitrary constants c_1 and
    c_2 as in Example 4. We write the general solution in the
    form $y = k_1 e^{(x-1)} + k_2 e^{-9(x-1)}$. Note that
    $c_1 = k_1 e^{-1}$ and $c_2 = k_2 e^9$. The advantage of the latter
    form of the general solution becomes clear when we apply
    the I.C. $y(1) = 1$ and $y'(1) = 0$. This latter form of y

gives $y' = k_1 e^{(x-1)} - 9k_2 e^{-9(x-1)}$ and thus setting $x = 1$ in y and y' yields the equations $k_1 + k_2 = 1$ and $k_1 - 9k_2 = 0$. Solving for k_1 and k_2 we find that $y = (9e^{(x-1)} + e^{-9(x-1)})/10$.

15. The general solution is $y = c_1 e^{-x} + c_2 e^{2x}$. Using the I.C. we obtain $c_1 + c_2 = \alpha$ and $-c_1 + 2c_2 = 2$, so adding the two equations we find $3c_2 = \alpha + 2$. If y is to approach zero as $x \to \infty$, c_2 must be zero. Thus $\alpha = -2$.

17. Let $v = y'$, then $v' = y''$ and thus the D.E. becomes $x^2 v' + 2xv - 1 = 0$ or $x^2 v' + 2xv = 1$. The left side is recognized as $(x^2 v)'$ and thus we may integrate to obtain $x^2 v = x + c$ (otherwise, divide both sides of the D.E. by x^2 and find the integrating factor, which is just x^2 in this case). Solving for $v = dy/dx$ we find $dy/dx = 1/x + c/x^2$ so that $y = \ln x + c_1/x + c_2$.

19. Set $v = y'$, then $v' = y''$ and thus the D.E. becomes $v' + xv^2 = 0$. This equation is separable and has the solution $-v^{-1} + x^2/2 = c$ or $v = y' = -2/(c_1 - x^2)$ where $c_1 = 2c$. We must consider separately the cases $c_1 = 0$, $c_1 > 0$ and $c_1 < 0$. If $c_1 = 0$, then $y' = 2/x^2$ or $y = -2/x + c_2$. If $c_1 > 0$, let $c_1 = k^2$. Then $y' = -2/(k^2-x^2) = -(1/k)[1/(k-x) + 1/(k+x)]$, so that $y = (1/k)\ln|(k-x)/(k+x)| + c_2$. If $c_1 < 0$, let $c_1 = -k^2$. Then $y' = 2/(k^2 + x^2)$ so that $y = (2/k)\tan^{-1}(x/k) + c_2$. Finally, we note that $y = $ constant is also a solution of the D.E.

23. Following the procedure outlined, let $v = dy/dx$ and $y'' = dv/dx = vdv/dy$. Thus the D.E. becomes $yvdv/dy + v^2 = 0$, which is a separable equation with the solution $v = c_1/y$. Next let $v = dy/dx = c/y$, which again separates to give the solution $y^2 = c_1 x + c_2$.

26. Again let $v = y'$ and $v' = vdv/dy$ to obtain $2y^2 vdv/dy + 2yv^2 = 1$. This is an exact equation with solution $v = \pm y^{-1}(y + c_1)^{1/2}$. To solve this equation,

we write it in the form $\pm y\,dy/(y+c_1)^{1/2} = dx$. On observing
that the left side of the equation can be written as
$\pm[(y+c_1) - c_1]dy/(y+c_1)^{1/2}$ we integrate and find
$\pm(2/3)(y-2c_1)(y+c_1)^{1/2} = x + c_2$.

28. If $v = y'$, then $v' = v\,dv/dy$ and the D.E. becomes
$v\,dv/dy + v^2 = 2e^{-y}$. Dividing by v we obtain
$dv/dy + v = 2v^{-1}e^{-y}$, which is a Bernoulli equation. Let
$w(y) = v^2$, then $dw/dy = 2v\,dv/dy$ and the last D.E.
becomes $dw/dy + 2w = 4e^{-y}$, which is linear in w. It's
solution is $w = v^2 = c_1e^{-2y} + 4e^{-y}$. Setting $v = dy/dx$, we
obtain a separable equation in y and x, which is solved
to yield the solution.

29. Since both x and y are missing, either approach used
above will work. In this case it's easier to use the
approach of Problems 15-20, so let $v = y'$ and thus $v' = y''$
and the D.E. becomes $v\,dv/dx = 2$.

32. The variable y is missing. Let $v = y'$, then $v' = y''$ and
the D.E. becomes $vv' - x = 0$. The solution of the
separable equation is $v^2 = x^2 + c_1$. Substituting $v = y'$ and
applying the I.C. $y'(1) = 1$, we obtain $y' = x$. The
positive square root was chosen because $y' > 0$ at $x = 1$.
Solving this last equation and applying the I.C. $y(1) = 2$,
we obtain $y = x^2/2 + 3/2$.

35. We have $y = c_1x + c_2\sin x$, $y' = c_1 + c_2\cos x$, $y'' = -c_2\sin x$.
From the last equation $c_2 = -y''/(\sin x)$. Substituting
this in the first equation gives $c_1 = (y+y'')/x$. Finally,
substituting for c_1 and c_2 in the second equation we
obtain $(1-x\cot x)y'' - xy' + y = 0$.

Section 3.2, Page 129

4. $W(x, xe^x) = \begin{vmatrix} x & xe^x \\ 1 & e^x + xe^x \end{vmatrix} = xe^x + x^2e^x - xe^x = x^2e^x$.

8. As in Example 1, $p(x) = -3x/(x-1)$ and $q(x) = 4/(x-1)$, so
the only point of discontinuity is $x = 1$. By Theorem
3.2.1, the largest interval is $-\infty < x < 1$, since the
initial point is $x_0 = -2$.

14. Substitute $y = 1$ into the D.E. to show that it is a solution. Similarly for $y = x^{1/2}$. If $y = c_1(1) + c_2x^{1/2}$ is substituted in the D.E. you will get $-c_1c_2/4x^{3/2}$, which is zero only if $c_1 = 0$ or $c_2 = 0$. Thus the linear combination of two solutions is not, in general, a solution. Theorem 3.2.2 is not contradicted however, since the D.E. is not linear.

15. $y = \phi(x)$ is a solution of the D.E. so $L[\phi](x) = g(x)$. Since L is a linear operator, $L[c\phi](x) = cL[\phi](x) = cg(x)$. But, since $g(x) \neq 0$, $cg(x) = g(x)$ if and only if $c = 1$. This is not a contradiction of Theorem 3.2.2 since the linear D.E. is not homogeneous.

18. $W(x,g) = \begin{vmatrix} x & g \\ 1 & g' \end{vmatrix} = xg' - g = x^2e^x$, or $g' - \frac{1}{x}g = xe^x$. This has an integrating factor of $\frac{1}{x}$ and thus $\frac{1}{x}g' - \frac{1}{x^2}g = e^x$ or $(\frac{1}{x}g)' = e^x$. Integrating and multiplying by x we obtain

$g(x) = xe^x + cx$.

21. From Section 3.1, e^x and e^{-2x} are two solutions, and since $W(e^x,e^{-2x}) \neq 0$ they form a fundamental set of solutions. To find the fundamental set specified by Theorem 3.2.5, let $y(x) = c_1e^x + c_2e^{-2x}$, where c_1 and c_2 satisfy $c_1 + c_2 = 1$ and $c_1 - 2c_2 = 0$ for y_1. Solving, we find $y_1 = \frac{2}{3}e^x + \frac{1}{3}e^{-2x}$. Likewise, c_1 and c_2 satisfy $c_1 + c_2 = 0$ and $c_1 - 2c_2 = 1$ for y_2, so that $y_2 = \frac{1}{3}e^x - \frac{1}{3}e^{-2x}$.

25. For $y_1 = x$, we have $x^2(0) - x(x+2)(1) + (x+2)(x) = 0$ and for $y_2 = xe^x$ we have $x^2(x+2)e^x - x(x+2)(x+1)e^x + (x+2)xe^x = 0$. From Problem 4, $W(x,xe^x) = x^2e^x \neq 0$ for $x > 0$, so y_1 and y_2 form a fundamental set of solutions.

27. Suppose that
     $P(x)y'' + Q(x)y' + R(x)y = [P(x)']' + [f(x)y]'$. On
     expanding the right side and equating coefficients, we
     find $f'(x) = R(x)$ and $P'(x) + f(x) = Q(x)$. These two
     conditions on f can be satisfied if
     $R(x) = Q'(x) - P''(x)$ which gives the necessary condition
     $P''(x) - Q'(x) + R(x) = 0$.

30. We have $P(x) = x$, $Q(x) = -\cos x$, and $R(x) = \sin x$ and the
     condition for exactness is satisfied. Also
     $f(x) = Q(x) - P'(x) = -\cos x - 1$, so the D.E. becomes
     $(xy')' - [(1 + \cos x)y]' = 0$. Hence
     $xy' - (1 + \cos x)y = c_1$. This is a first order linear
     D.E. and the integrating factor (after dividing by x) is
     $\mu(x) = \exp[-\int x^{-1}(1 + \cos x)dx]$. The general solution is
     $$y = [\mu(x)]^{-1}[c_1\int_{x_0}^{x} t^{-1} \mu(t)dt + c_2].$$

32. We want to choose $\mu(x)$ and $f(x)$ so that $\mu(x)P(x)y'' +$
     $\mu(x)Q(x)y' + \mu(x)R(x)y = [\mu(x)P(x)y']' + [f(x)y]'$.
     Expand the right side and equate coefficients of y'', y'
     and y. This gives $\mu'(x)P(x) + \mu(x)P'(x) + f(x) =$
     $\mu(x)Q(x)$ and $f'(x) = \mu(x)R(x)$. Differentiate the first
     equation and then eliminate $f'(x)$ to obtain the adjoint
     equation $P\mu'' + (2P' - Q)\mu' + (P'' - Q' + R)\mu = 0$.

36. Write the adjoint D.E. given in Problem 32 as
     $\hat{P}\mu'' + \hat{Q}\mu' + \hat{R}\mu = 0$ where $\hat{P} = P$, $\hat{Q} = 2P' - Q$, and
     $\hat{R} = P'' - Q' + R$. The adjoint of this equation, namely
     the adjoint of the adjoint, is
     $\hat{P}y'' + (2\hat{P}' - \hat{Q})y' + (\hat{P}'' - \hat{Q}' + \hat{R})y = 0$. After
     substituting for $\hat{P}$, $\hat{Q}$, and $\hat{R}$ and simplifying, we obtain
     $Py'' + Qy' + Ry = 0$. This is the same as the original
     equation.

37. From Problem 32 the adjoint of $Py'' + Qy' + Ry = 0$ is
     $P\mu'' + (2P' - Q)\mu' + (P'' - Q' + R)\mu = 0$. The two equations
     are the same if $2P' - Q = Q$ and $P'' - Q' + R = R$. This
     will be true if $P' = Q$. Hence the original D.E. is self-
     adjoint if $P' = Q$. For Problem 33, $P(x) = x^2$ so
     $P'(x) = 2x$ and $Q(x) = x$. Hence the Bessel equation of
     order ν is not self-adjoint. In a similar manner we find
     that Problems 34 and 35 are self-adjoint.

Section 3.3, Page 135

2.   Since $\cos 3x = 4\cos^3 x - 3\cos x$ we have
      $\cos 3x - (4\cos^3 x - 3\cos x) = 0$ for all x. From Eq.(1) we
      have $k_1 = 1$ and $k_2 = -1$ and thus $\cos 3x$ and $4\cos^3 x - 3\cos x$
      are linearly dependent.

6. $W(x, x^{-1}) = -2/x \neq 0$.

10.   The D.E. is linear and homogeneous. Hence, if y_1 and y_2
      are solutions, then $y_3 = y_1 + y_2$ and $y_4 = y_1 - y_2$ are
      solutions. $W(y_3, y_4) = y_3 y_4' - y_3' y_4 = (y_1 + y_2)(y_1' - y_2') -$
      $(y_1' + y_2')(y_1 - y_2) = -2(y_1 y_2' - y_1' y_2) = -2W(y_1, y_2)$, is not
      zero since y_1 and y_2 are linearly independent solutions.
      Hence y_3 and y_4 form a fundamental set of solutions.

13.   Writing the D.E. in the form of Eq.(6), we have
      $p(x) = -(x+2)/x$. Thus Eq.(7) yields
      $$W(x) = c \exp[-\int \frac{-(x+2)}{x} dx] = cx^2 e^x.$$

16.   From Eq.(7) we have $W(y_1, y_2) = c \exp[-\int p(x)dx]$, where
      $p(x) = 2/x$ from the D.E. Thus $W(y_1, y_2) = c/x^2$. Since
      $W(y_1, y_2)(1) = 2$ we find $c = 2$ and thus $W(y_1, y_2)(5) = 2/25$.

19.   Let c be in a point in I at which both y_1 and y_2 vanish.
      Then $W(y_1, y_2)(c) = y_1(c)y_2'(c) - y_1'(c)y_2(c) = 0$. Hence, by
      Theorem 3.3.3 the functions y_1 and y_2 cannot form a
      fundamental set.

21.   Suppose that y_1 and y_2 have a point of inflection at x_0
      and either $p(x_0) \neq 0$ or $q(x_0) \neq 0$. Since $y_1''(x_0) = 0$ and
      $y_2''(x_0) = 0$ it follows from the D.E. that $p(x_0)y_1'(x_0) +$
      $q(x_0)y_1(x_0) = 0$ and $p(x_0)y_2'(x_0) + q(x_0)y_2(x_0) = 0$. If
      $p(x_0) = 0$ and $q(x_0) \neq 0$ then $y_1(x_0) = y_2(x_0) = 0$, and
      $W(y_1, y_2)(x_0) = 0$ so the solutions cannot form a
      fundamental set. If $p(x_0) \neq 0$ and $q(x_0) = 0$ then
      $y_1'(x_0) = y_2'(x_0) = 0$ and $W(y_1, y_2)(x_0) = 0$, so again the

solutions cannot form a fundamental set. If $p(x_0) \neq 0$
and $q(x_0) = 0$ then $y_1'(x_0) = q(x_0)y_1(x_0)/p(x_0)$ and
$y_2'(x_0) = q(x_0)y_2(x_0)/p(x_0)$ and thus

$$
\begin{aligned}
W(y_1, y_2)(x_0) &= y_1(x_0)y_2'(x_0) - y_1'(x_0)y_2(x_0) \\
&= y_1(x_0)[q(x_0)y_2(x_0)/p(x_0)] - \\
&\quad [q(x_0)y_1(x_0)/p(x_0)]y_2(x_0) \\
&= 0.
\end{aligned}
$$

22. Let $-1 < x_0, x_1 < 1$ and $x_0 \neq x_1$. If $y_1 = x$ and $y_2 = x^2$ are
linearly dependent then $c_1x_1 + c_2x_1^2 = 0$ and $c_1x_0 + c_2x_0^2 = 0$
have a solution for c_1 amd c_2 such that c_1 and c_2 are not
both zero. But this system of equations has a non-zero
solution only if $x_1 = 0$ or $x_0 = 0$ or $x_1 = x_0$. Hence, the
only set c_1 and c_2 that satisfies the system for every
choice of x_0 and x_1 in $-1 < x < 1$ is $c_1 = c_2 = 0$.
Therefore x and x^2 are linearly independent on
$-1 < x < 1$. Next, $W(x, x^2) = x^2$ clearly vanishes at $x = 0$.
Since $W(x, x^2)$ vanishes at $x = 0$, but x and x^2 are
linearly independent on $-1 < x < 1$, it follows that x and
x^2 cannot be solutions of Eq.(6) on $-1 < x < 1$. To show
that the functions $y_1 = x$ and $y_2 = x^2$ are solutions of
$x^2y'' - 2xy' + 2y = 0$, substitute each of them in the
equation. Clearly, they are solutions. There is no
contradiction to Theorem 3.3.3 since $p(x) = -2/x$ and
$q(x) = 2/x^2$ are discontinuous at $x = 0$, and hence the
theorem does not apply on the interval $-1 < x < 1$.

23. On $0 < x < 1$, $y_1(x) = x^2$ and $y_2(x) = x^2$. Hence there are
nonzero constants, $c_1 = 1$ and $c_2 = -1$, such that
$c_1y_1(x) + c_2y_2(x) = 0$ for each x in $(0,1)$. On $-1 < x < 0$,
$y_1(x) = -x^2$ and $y_2(x) = x^2$; thus $c_1 = c_2 = 1$ defines
constants such that $c_1y_1(x) + c_2y_2(x) = 0$ for each x in
$(-1,0)$. Thus y_1 and y_2 are linearly dependent on
$0 < x < 1$ and on $-1 < x < 0$. We will show that $y_1(x)$ and
$y_2(x)$ are linearly independent on $-1 < x < 1$ by
demonstrating that it is impossible to find constants c_1
and c_2, not both zero, such that $c_1y_1(x) + c_2y_2(x) = 0$ for
all x in $(-1,1)$. Assume that there are two such nonzero

constants and choose two points x_0 and x_1 in $-1 < x < 1$
such that $x_0 < 0$ and $x_1 > 0$. Then $-c_1x_0^2 + c_2x_0^2 = 0$ and
$c_1x_1^2 + c_2x_1^2 = 0$. These equations have a nontrivial
solution for c_1 and c_2 only if the determinant of
coefficients is zero. But the determinant of
coefficients is $-2x_0^2x_1^2 \neq 0$ for x_0 and x_1 as specified.
Hence $y_1(x)$ and $y_2(x)$ are linearly independent on
$-1 < x < 1$.

Section 3.4, Page 142

1. $\exp(1+2i) = e^{1+2i} = ee^{2i} = e(\cos 2 + i \sin 2)$.

5. Recall that $2^{1-i} = e^{\ln(2^{1-i})} = e^{(1-i)\ln 2}$.

7. As in Section 3.1, we seek solutions of the form $y = e^{rx}$.
    Substituting this into the D.E. yields the characteristic
    equation $r^2 - 2r + 2 = 0$, which has the roots $r_1 = 1 + i$
    and $r_2 = 1 - i$, using the quadratic formula. Thus $\lambda = 1$
    and $\mu = 1$ and from Eq.(13) the general solution is
    $y = c_1e^x\cos x + c_2e^x\sin x$.

14. The characteristic equation is $9r^2 + 9r - 4$, which has
    the real roots $-4/3$ and $1/3$. Thus the solution has the
    same form as in Section 3.1, $y(x) = c_1e^{x/3} + c_2e^{-4/3}$.

18. The characteristic equation is
    $r^2 + 4r + 5 = 0$, which has the
    roots $r_1, r_2 = -2 \pm i$. Thus
    $y = c_1e^{-2x}\cos x + c_2e^{-2x}\sin x$ and
    $y' = (-2c_1+c_2)e^{-2x}\cos x +$
    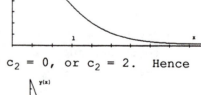
    $(-c_1-2c_2)e^{-2x}\sin x$, so that
    $y(0) = c_1 = 1$ and $y'(0) = -2c_1 + c_2 = 0$, or $c_2 = 2$. Hence
    $y = e^{-2x}(\cos x + 2\sin x)$.

22. The characteristic equation is
    $r^2 + 2r + 2 = 0$, so $r_1, r_2 = -1 \pm i$.
    Since the I.C. are given at $\pi/4$
    we assume $y = e^{-(x-\pi/4)}(c_1\cos x + c_2\sin x)$

so $y' = -e^{-(x-\pi/4)}(c_1\cos x + c_2\sin x) +$
$e^{-(x-\pi/4)}(-c_1\sin x + c_2\cos x)$. Thus
$\sqrt{2}\,c_1/2 + \sqrt{2}\,c_2/2 = 2$ and $-\sqrt{2}\,c_1 = -2$ and
hence $y = \sqrt{2}\,e^{-(x-\pi/4)}(\cos x + \sin x)$.

27. Suppose that $x = a$ and $x = b$ ($b>a$) are consecutive zeros
of y_1. We must show that y_2 vanishes once and only once
in the interval $a < x < b$. Assume that it does not
vanish. Then we can form the quotient y_1/y_2 on the
interval $a \leq x \leq b$. Note $y_2(a) \neq 0$ and $y_2(b) \neq 0$,
otherwise y_1 and y_2 would not be linearly independent
solutions. Next, y_1/y_2 vanishes at $x = a$ and $x = b$ and
has a derivative in $a < x < b$. By Rolles theorem, the
derivative must vanish at an interior point. But

$$\left(\frac{y_1}{y_2}\right)' = \frac{y_1'y_2 - y_2'y_1}{y_2^2} = \frac{-W(y_1, y_2)}{y_2^2}, \text{ which cannot be zero}$$

since y_1 and y_2 are linearly independent solutions.
Hence we have a contradiction, and we conclude that y_2
must vanish at a point between a and b. Finally, we show
that it can vanish at only one point between a and b.
Suppose that it vanishes at two points c and d between a
and b. By the argument we have just given we can show
that y_1 must vanish between c and d. But this contradicts
the hypothesis that a and b are consecutive zeros of y_1.

29. We use the result of Problem 28. Note that $q(x) = e^{-x^2} > 0$
for $-\infty < x < \infty$. Next, we find that $(q' + 2pq)/q^{3/2} = 0$.
Hence the D.E. can be transformed into an equation with
constant coefficients by letting $z = u(x) = \int e^{-x^2/2}dx$.
Substituting $z = u(x)$ in the differential equation found
in part (b) of Problem 28 we obtain, after dividing by
the coefficient of d^2y/dz^2, the D.E. $d^2y/dz^2 - y = 0$.
Hence the general solution of the original D.E. is
$y = c_1\cos z + c_2\sin z$, $z = \int e^{-x^2/2}dx$.

32. Rewrite the D.E. as $y'' + (\alpha/x)y' + (\beta/x^2)y = 0$ so that
$p = \alpha/x$ and $q = \beta/x^2$, which satisfy the conditions of
parts (c) and (d) of Problem 28. Thus
$z = \int (1/x^2)^{1/2}dx = \ln x$ will transform the D.E. into
$dy^2/dz^2 + (\alpha-1)\,dy/dz + \beta y = 0$. Note that since β is
constant, it can be neglected in defining z.

33. By direct substitution or from Problem 32, $z = \ln x$ will
     transform the D.E. into $d^2y/dz^2 + y = 0$, since $\alpha = 1$ and
     $\beta = 1$. Thus $y = c_1\cos z + c_2\sin z$, with $z = \ln x$, $x > 0$.

Section 3.5, Page 150

1.   Substituting $y = e^{rx}$ into the D.E., we find that
     $r^2 - 2r + 1 = 0$, which gives $r_1 = 1$ and $r_2 = 1$. Since the
     roots are equal, the second linearly independent solution
     is xe^x and thus the general solution is $y = c_1e^x + c_2xe^x$.

9.   The characteristic equation is $25r^2 - 20r + 4 = 0$, which
     may be written as $(5r-2)^2 = 0$ and hence the roots are
     $r_1, r_2 = 2/5$. Thus $y = c_1e^{2x/5} + c_2\, xe^{2x/5}$.

14. The characteristic equation is
     $r^2 + 4r + 4 = (r+2)^2 = 0$, which
     has the repeated root $r = -2$.
     Since the I.C. are given at
     $x = -1$, write the general solution
     as $y = c_1e^{-2(x+1)} + c_2xe^{-2(x+1)}$.   Then
     $y' = -2c_1e^{-2(x+1)} + c_2e^{-2(x+1)} - 2c_2xe^{-2(x+1)}$
     and hence $c_1-c_2 = 2$ and $-2c_1+3c_2 = 1$
     which yield $c_1 = 7$ and $c_2 = 5$.   Thus
     $y = 7e^{-2(x+1)} + 5xe^{-2(x+1)}$, a decaying exponential as shown
     in the graph.

16. If $r_2 \neq r_1$ then $\phi(x; r_1, r_2) = (e^{r_2x} - e^{r_1x})/(r_2 - r_1)$ is
     defined for all x. Note that ϕ is a linear combination
     of two solutions, e^{r_1x} and e^{r_2x}, of the D.E. Hence, ϕ is a
     solution of the differential equation. Think of r_1 as
     fixed and let $r_2 \to r_1$. The limit of ϕ as $r_2 \to r_1$ is
     indeterminate. If we use L'Hopital's rule, we find
     $$\lim_{r_2 \to r_1} \frac{e^{r_2x} - e^{r_1x}}{r_2 - r_1} = \lim_{r_2 \to r_1} \frac{xe^{r_2x}}{1} = xe^{r_1x}. \quad \text{Hence, the}$$
     solution $\phi(x; r_1, r_2) \to xe^{r_1x}$ as $r_2 \to r_1$.

18. As in Eq.(25) let $y_2 = v\cdot 1 = v$. Then $y_2' = v'$, $y_2'' = v''$ and
     on substituting in the D.E. we obtain
     $x^2v'' + 2xv' = (x^2v')' = 0$. Hence $v' = c_1x^{-2}$ and

$v = -c_1 x^{-1} + c_2$, so $y_2 = (-c_1 x^{-1} + c_2)$. The constant c_2 adds only a multiple of y_1 and we can take $c_1 = -1$, so the second linearly independent solution is $y_2(x) = x^{-1}$.

20. Let $y_2 = v/x$. Then $y_2' = v'/x - v/x^2$ and $y_2'' = v''/x - 2v'/x^2 + 2v/x^3$. Substituting in the D.E. we obtain $x^2(v''/x - 2v'/x^2 + 2v/x^3) + 3x(v'/x - v/x^2) + v/x = 0$. Simplifying the left side we get $xv'' + v' = 0$, which yields $v' = c_1/x$. Thus $v = c_1 \ln x + c_2$. Hence a second solution is $y_2(x) = (c_1 \ln x + c_2)/x$. However, we may set $c_2 = 0$ and $c_1 = 1$ without loss of generality and thus we have $y_2(x) = (\ln x)/x$ as a second solution. Note that in the form we actually calculated, $y_2(x)$ is a linear combination of $1/x$ and $\ln x/x$, and hence is the general solution.

22. In this case the calculations are somewhat easier if we do not use the explicit form for $y_1(x) = \sin x^2$ at the beginning but simply set $y_2(x) = y_1 v$. Substituting this form for y_2 in the D.E. gives $x(y_1 v)'' - (y_1 v)' + 4x^3(y_1 v) = 0$. On carrying out the differentiations and making use of the fact that y_1 is a solution, we obtain $xy_1 v'' + (2xy_1' - y_1)v' = 0$. This is a first order linear equation for v', which has the solution $v' = cx/(\sin x^2)^2$. Setting $u = x^2$ allows integration of this to get $v = c_1 \cot x^2 + c_2$. Setting $c_1 = 1$, $c_2 = 0$ and multiplying by $y_1 = \sin x^2$ we obtain $y_2(x) = \cos x^2$ as the second solution of the D.E.

25. Substituting $y_2(x) = y_1(x) v(x)$ in the D.E. gives $x^2(y_1 v)'' + x(y_1 v)' + (x^2 - \frac{1}{4})y_1 v = 0$. On carrying out the differentiations and making use of the fact that y_1 is a solution, we obtain $x^2 y_1 v'' + (2x^2 y_1' + xy_1)v' = 0$. This is a first order linear equation for v', $v'' + (2y_1'/y_1 + 1/x)v' = 0$, with solution

$$v'(x) = c\exp[-\int(2\frac{y_1'}{y_1} + \frac{1}{x})dx] = c\exp[-2\ln y_1 - \ln x]\ \text{or}$$

$$v' = c\frac{1}{xy_1{}^2} = \frac{c}{x(x^{-1}\sin^2 x)} = c\csc^2 x,$$

where c is an arbitrary constant, which we will take to be one. Then $v(x) = \int\csc^2 x\,dx = -\cot x + k$ where again k is an arbitrary constant which can be taken equal to zero. Thus

$y_2(x) = y_1(x)v(x) = (x^{-1/2}\sin x)(-\cot x) = x^{-1/2}\cos x.$ The minus sign is unimportant since the solution of a homogeneous equation can be multiplied by any constant. The second solution is usually taken to be $x^{-1/2}\cos x$. Note that $c = -1$ would have given this solution.

26b. Let $y_2(x) = e^x v(x)$, then $y_2' = e^x v' + e^x v$, and

$y_2'' = e^x v'' + 2e^x v' + e^x v.$ Substituting in the D.E. we obtain $xe^x v'' + (xe^x - Ne^x)v' = 0$, or $v'' + (1-N/x)v' = 0$. This is a first order linear D.E. for v' with integrating factor $\mu(x) = \exp[\int(1-N/x)dx] = x^{-N}e^x$. Hence $(x^{-N}e^x v')' = 0$, and $v' = cx^N e^{-x}$ which gives $v(x) = c\int x^N e^{-x}dx + k.$ On taking $k = 0$ we obtain as the second solution $y_2(x) = ce^x\int x^N e^{-x}dx.$ The integral can be evaluated by using the method of integration by parts. At each stage let $u = x^N$ or x^{N-1}, or whatever the power of x that remains, and let $dv = e^{-x}$. Note that this dv is not related to the $v(x)$ in $y_2(x)$. For $N = 2$ we have

$$y_2(x) = ce^x\int x^2 e^{-x}dx = ce^x[x^2\frac{e^{-x}}{-1} - \int 2x\frac{e^{-x}}{-1}dx]$$

$$= -cx^2 + ce^x[2x\frac{e^{-x}}{-1} - \int 2\frac{e^{-x}}{-1}dx]$$

$$= c(-x^2 - 2x - 2) = -2c(1 + x + x^2/2!).$$

Choosing $c = -1/2!$ gives the desired result. For the general case $c = -1/N!$

28. $(y_2/y_1)' = (y_1 y_2' - y_1' y_2)/y_1{}^2 = W(y_1, y_2)/y_1{}^2.$ Abel's identity is $W(y_1, y_2) = c\exp[-\int p(x)dx].$ Hence $(y_2/y_1)' = cy_1{}^{-2}\exp[-\int p(x)dx].$ Integrating and setting $c = 1$ (since a solution y_2 can be multiplied by any constant) and taking the constant of integration to be zero we obtain

$$y_2(x) = y_1(x)\int\frac{\exp[-\int p(x)dx]}{[y_1(x)]^2}\,dx.$$

30. From Problem 28 and Abel's formula we have

$$(\frac{y_2}{y_1})' = \frac{\exp[\int(1/x)\,dx]}{\sin^2(x^2)} = \frac{e^{\ln x}}{\sin^2(x^2)} = x\csc^2(x^2).$$ Thus

$y_2/y_1 = -(1/2)\cot(x^2)$ and hence we can choose $y_2 = \cos(x^2)$

since $y_1 = \sin^2(x^2)$.

33. The general solution of the D.E. is $y = c_1 e^{r_1 x} + c_2 e^{r_2 x}$

where $r_1, r_2 = (-b \pm \sqrt{b^2 - 4ac})/2a$ provided $b^2 - 4ac \neq 0$.

In this case there are two possibilities. If $b^2 - 4ac > 0$

then $(b^2 - 4ac)^{1/2} < b$ and r_1 and r_2 are real and

negative. Consequently $e^{r_1 x} \rightarrow 0$ and $e^{r_2 x} \rightarrow 0$; and hence

$y \rightarrow 0$, as $x \rightarrow \infty$. If $b^2 - 4ac < 0$ then r_1 and r_2 are

complex conjugates with negative real part. Again

$e^{r_1 x} \rightarrow 0$ and $e^{r_2 x} \rightarrow 0$; and hence $y \rightarrow 0$, as $x \rightarrow \infty$.

Finally, if $b^2 - 4ac = 0$, then $y = c_1 e^{r_1 x} + c_2 x e^{r_1 x}$ where

$r_1 = -b/2a < 0$. Hence, again $y \rightarrow 0$ as $x \rightarrow \infty$. This

conclusion does not hold if either $b = 0$ or $c = 0$.

37. Substituting $z = \ln x$ into the D.E. gives

$$\frac{d^2 y}{dz^2} + \frac{dy}{dz} + 0.25y = 0,$$ which has the solution

$y(z) = c_1 e^{-z/2} + c_2 z e^{-z/2}$ so that $y(x) = c_1 x^{-1/2} + c_2 x^{-1/2}\ln x$.

Section 3.6, Page 162

1. First we find the solution of the homogeneous D.E., which
    has the characteristic equation $r^2 - 2r - 3 = (r-3)(r+1) = 0$.
    Hence $y_c = c_1 e^{3x} + c_2 e^{-x}$ and we can assume $Y = A e^{2x}$ for the
    particular solution. Thus $Y' = 2A e^{2x}$ and $Y'' = 4A e^{2x}$ and
    substituting into the D.E. yields
    $4A e^{2x} - 2(2A e^{2x}) - 3(A e^{2x}) = 3e^{2x}$. Thus $-3A = 3$ and
    $A = -1$, yielding $y = c_1 e^{3x} + c_c e^{-x} - e^{2x}$.

4. Initially we assume $Y = A + B\sin 2x + C\cos 2x$. However,
    since a constant is a solution of the related homogeneous
    D.E. we must modify Y by multiplying the constant A by x
    and thus the correct form is $Y = Ax + B\sin 2x + C\cos 2x$.

6. Since $y_c = c_1e^{-x} + c_2xe^{-x}$ we must assume $Y = Ax^2e^{-x}$, so that
    $Y' = 2Axe^{-x} - Ax^2e^{-x}$ and $y'' = 2Ae^{-x} - 4Axe^{-x} + Ax^2e^{-x}$.
    Substituting in the D.E. gives $(Ax^2-4Ax+2A)e^{-x} +$
    $2(-Ax^2+2Ax)e^{-x} + Ax^2e^{-x} = 2e^{-x}$. Notice that all terms on
    the left involving x^2 and x add to zero and we are left
    with $2A = 2$, or $A = 1$. Hence $y = c_1e^{-x} + c_2xe^{-x} + x^2e^{-x}$.

8. The assumed form is $Y = (Ax + B)\sin 2x + (Cx + D)\cos 2x$,
    which is appropriate for both terms appearing on the
    right side of the D.E. Since none of the terms appearing
    in Y are solutions of the homogeneous equation, we do not
    need to modify Y.

11. First solve the homogeneous D.E. Substituting $y = e^{rx}$
    gives $r^2 + r + 4 = 0$. Hence $y_c = e^{-x/2}[c_1\cos(\sqrt{15}\,x/2) +$
    $c_2\sin(\sqrt{15}\,x/2)]$. We replace sinhx by $(e^x - e^{-x})/2$ and
    then assume $Y(x) = Ae^x + Be^{-x}$. Since neither e^x nor e^{-x} are
    solutions of the homogeneous equation, there is no need
    to modify our assumption for Y. Substituting in the D.E.,
    we obtain $6Ae^x + 4Be^{-x} = e^x - e^{-x}$. Hence, $A = 1/6$ and
    $B = -1/4$. The general solution is $y = e^{-x/2}[c_1\cos(\sqrt{15}\,x/2)$
    $+ c_2\sin(\sqrt{15}\,x/2)] + e^x/6 - e^{-x}/4$. [For this problem we
    could also have found a particular solution as a linear
    combination of sinhx and coshx: $Y(x) = A\cosh x + B\sinh x$.
    Substituting this in the D.E. gives
    $(5A + B)\cosh x + (A + 5B)\sinh x = 2\sinh x$. The solution is
    $A = -1/12$ and $B = 5/12$. A simple calculation shows that
    $-(1/12)\cosh x + (5/12)\sinh x = e^x/6 - e^{-x}/4$.]

13. $y_c = c_1e^{-2x} + c_2e^x$ so for the particular solution we assume
    $Y = Ax + B$. Since neither Ax or B are solutions of the
    homogeneous equation it is not necessary to modify the
    original assumption. Substituting Y in the D.E. we
    obtain $0 + A - 2(Ax+B) = 2x$ or $-2A = 2$ and $A-2B = 0$.
    Solving for A and B we obtain $y = c_1e^{-2x} + c_2e^x - x - 1/2$
    as the general solution. $y(0) = 0 \rightarrow c_1 + c_2 - 1/2 = 0$
    and $y'(0) = 1 \rightarrow -2c_1 + c_2 - 1 = 1$, which yield $c_1 = -1/2$
    and $c_2 = 1$. Thus $y = e^x - (1/2)e^{-2x} - x - 1/2$.

16. Since the nonhomogeneous term is the product of a linear
    polynomial and an exponential, assume Y of the same form:

$Y = (Ax+B)e^{2x}$.

19. The solution of the homogeneous D.E. is $y_c = c_1 e^{-3x} + c_2$. After inspection of the nonhomogeneous term, we assume $Y(x) = (A_0 x^4 + A_1 x^3 + A_2 x^2 + A_3 x + A_4) + (B_0 x^2 + B_1 x + B_2)e^{-3x} +$ $C \sin 3x + D \cos 3x$. However, since e^{-3x} and a constant are solutions of the homogeneous D.E., we must multiply the coefficient of e^{-3x} and the polynomial by x. The correct form is $Y(x) = x(A_0 x^4 + A_1 x^3 + A_2 x^2 + A_3 x + A_4) + x(B_0 x^2 + B_1 x + B_2)e^{-3x} + C \sin 3x + D \cos 3x$.

22. The solution of the homogeneous D.E. is $y_c = e^{-x}[c_1 \cos x + c_2 \sin x]$. After inspection of the nonhomogeneous term, we assume $Y(x) = Ae^{-x} + (B_0 x^2 + B_1 x + B_2)e^{-x} \cos x + (C_0 x^2 + C_1 x + C_2)e^{-x} \sin x$. Since $e^{-x} \cos x$ and $e^{-x} \sin x$ are solutions of the homogeneous D.E., it is necessary to multiply both of these terms by x. Hence the correct form is $Y(x) = Ae^{-x} + x(B_0 x^2 + B_1 x + B_2)e^{-x} \cos x + x(C_0 x^2 + C_1 x + C_2)e^{-x} \sin x$.

28. First solve the I.V.P. $y'' + y = t$, $y(0) = 0$, $y'(0) = 1$ for $0 \le t \le \pi$. The solution of the homogeneous D.E. is $y_c(t) = c_1 \cos t + c_2 \sin t$. The correct form for $Y(t)$ is $Y(t) = A_0 t + A_1$. Substituting in the D.E. we find $A_0 = 1$ and $A_1 = 0$. Hence, $y = c_1 \cos t + c_2 \sin t + t$. Applying the I.C., we obtain $y = t$. For $t > \pi$ the complementary solution is $y_c(t) = D_1 \cos t + D_2 \sin t$ and the form for $Y(t)$ is $Y(t) = Ee^{\pi-t}$. Substituting $Y(t)$ in the D.E., we obtain $Ee^{\pi-t} + Ee^{\pi-t} = \pi e^{\pi-t}$ so $E = \pi/2$. Hence the general solution for $t > \pi$ is $Y = D_1 \cos t + D_2 \sin t + (\pi/2)e^{\pi-t}$. If y and y' are to be continuous at $t = \pi$, then the solutions and their derivatives for $t \le \pi$ and $t > \pi$ must have the same value at $t = \pi$. These conditions require $\pi = -D_1 + \pi/2$ and $1 = -D_2 - \pi/2$. Hence $D_1 = -\pi/2$, $D_2 = -(1 + \pi/2)$, and

$$y = \phi(t) = \begin{cases} t, & 0 \le t \le \pi \\ -(\pi/2)\cos t - (1 + \pi/2)\sin t + (\pi/2)e^{\pi-t}, & t > \pi. \end{cases}$$

The graphs are shown on the next page.

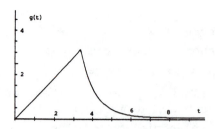
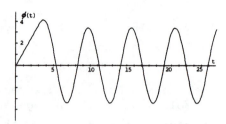

30. According to Theorem 3.6.1, the difference of any two
    solutions of the linear second order nonhomogeneous D.E.
    is a solution of the corresponding homogeneous D.E.
    Hence $y_1 - y_2$ is a solution of $ay'' + by' + cy = 0$. In
    Problem 33 of Section 3.5 we showed that if $a > 0$, $b > 0$,
    and $c > 0$ then every solution of this D.E. goes to zero
    as $t \to \infty$.

33. From Problem 32 we write the D.E. as $(D-4)(D+1)y = 3e^{2x}$.
    Thus let $(D+1)y = u$ and then $(D-4)u = 3e^{2x}$. This last
    equation is the same as $du/dx - 4u = 3e^{2x}$, which may be
    solved by multiplying both sides by e^{-4x} and integrating
    (see section 2.1). This yields $u = (-3/2)e^{2x} + Ce^{4x}$.
    Substituting this form of u into $(D+1)y = u$ we obtain
    $dy/dx + y = (-3/2)e^{2x} + Ce^{4x}$. Again, multiplying by e^x
    and integrating gives $y = (-1/2)e^{2x} + C_1e^{4x} + C_2e^{-x}$, where
    $C_1 = C/5$.

Section 3.7, Page 168

2. Two linearly independent solutions of the homogeneous
   D.E. are $y_1(x) = e^{2x}$ and $y_2(x) = e^{-x}$. Assume $Y = u_1(x)e^{2x} +$
   $u_2(x)e^{-x}$, then $Y'(x) = [2u_1(x)e^{2x} - u_2(x)e^{-x}] + [u_1'(x)e^{2x} +$
   $u_2'(x)e^{-x}]$. We set $u_1'(x)e^{2x} + u_2'(x)e^{-x} = 0$. Computing $Y''(x)$
   and substituting in the D.E. gives
   $2u_1'(x)e^{2x} - u_2'(x)e^{-x} = 2e^{-x}$. Thus we have two algebraic
   equations for $u_1'(x)$ and $u_2'(x)$ with the solution
   $u_1'(x) = 2e^{-3x}/3$ and $u_2'(x) = -2/3$. Hence $u_1(x) = -2e^{-3x}/9$
   and $u_2(x) = -2x/3$. Substituting in the formula for $Y(x)$
   we obtain $Y(x) = (-2e^{-3x}/9)e^{2x} + (-2x/3)e^{-x} = (-2e^{-x}/9) -$
   $(2xe^{-x}/3)$. Since e^{-x} is a solution of the homogeneous

D.E., we can choose $Y(x) = -2xe^{-x}/3$.

5. Since $\cos x$ and $\sin x$ are solutions of the homogeneous
    D.E., we assume $Y = u_1(x)\cos x + u_2(x)\sin x$. Thus
    $y' = -u_1(x)\sin x + u_2(x)\cos x$ after setting
    $u_1'(x)\cos x + u_2'(x)\sin x = 0$. Finding Y'' and substituting
    into the D.E. then yields $-u_1'(x)\sin x + u_2'(x)\cos x = \tan x$.
    The two equations for $u_1'(x)$ and $u_2'(x)$ have the solution
    $u_1'(x) = -\sin^2 x/\cos x = -\sec x + \cos x$ and
    $u_2'(x) = \sin x$. Thus $u_1(x) = \sin x - \ln(\tan x + \sec x)$ and
    $u_2(x) = -\cos x$, which when substituted into the assumed
    form for y, simplified, and added to the homogeneous
    solution yields
    $y = c_1\cos x + c_2\sin x - (\cos x)\ln(\tan x + \sec x)$.

11. Two linearly independent solutions of the homogeneous
    D.E. are $y_1(x) = e^{3x}$ and $y_2(x) = e^{2x}$. Applying Theorem
    3.7.1 with $W(y_1, y_2)(x) = -e^{5x}$, we obtain
    $$Y(x) = -e^{3x}\int\frac{e^{2t}g(t)}{-e^{5t}}\, dt + e^{2x}\int\frac{e^{3t}g(t)}{-e^{5t}}\,dt$$
    $$= \int [e^{3(x-t)} - e^{2(x-t)}]g(t)\,dt.$$

    The complete solution is then obtained by adding
    $c_1e^{3x} + c_2e^{2x}$ to $Y(x)$.

14. That x and xe^x are solutions of the homogeneous D.E. can
    be verified by direction substitution. Thus we assume
    $Y = xu_1(x) + xe^xu_2(x)$. Following the pattern of earlier
    problems we find $xu_1'(x) + xe^xu_2'(x) = 0$ and
    $u_1'(x) + (x+1)e^xu_2' = 2x$. [Note that $g(x) = 2x$, since the
    D.E. must be put into the form of Eq.(16)]. The solution
    of these equations gives $u_1'(x) = -2$ and $u_2'(x) = 2e^{-x}$.
    Hence, $u_1(x) = -2x$ and $u_2(x) = -2e^{-x}$, and
    $Y(x) = x(-2x) + xe^x(-2e^{-x}) = -2x^2 - 2x$. However, since x
    is a solution of the homogeneous D.E. we can choose as
    our particular solution $Y(x) = -2x^2$.

18. For this problem, and for many others, it is probably

easier to rederive Eqs.(26) without using the explicit
form for $y_1(x)$ and $y_2(x)$ and then to substitute for $y_1(x)$
and $y_2(x)$ in Eqs.(26). In this case if we take
$y_1 = x^{-1/2}\sin x$ and $y_2 = x^{-1/2}\cos x$, then $W(y_1,y_2) = -1/x$. If
the D.E. is put in the form of Eq.(16), then
$g(x) = 3x^{-1/2}\sin x$ and thus $u_1'(x) = 3\sin x\cos x$ and
$u_2'(x) = -3\sin^2 x = 3(-1 + \cos 2x)/2$. Hence
$u_1(x) = (3\sin^2 x)/2$ and $u_2(x) = -3x/2 + 3(\sin 2x)/4$, and

$$Y(x) = \frac{3\sin^2 x}{2}\frac{\sin x}{\sqrt{x}} + \left(-\frac{3x}{2} + \frac{3\sin 2x}{4}\right)\frac{\cos x}{\sqrt{x}}$$

$$= \frac{3\sin^2 x}{2}\frac{\sin x}{\sqrt{x}} + \left(-\frac{3x}{2} + \frac{3\sin x\cos x}{2}\right)\frac{\cos x}{\sqrt{x}}$$

$$= \frac{3\sin x}{2\sqrt{x}} - \frac{3\sqrt{x}\cos x}{2}.$$

The first term is a multiple of $y_1(x)$ and thus can be
neglected for $Y(x)$.

22. Putting limits on the integrals of Eq.(28) and changing
the integration variable to t yields

$$Y(x) = -y_1(x)\int_{x_0}^{x}\frac{y_2(t)g(t)dt}{W(y_1,y_2)(t)} + y_2(x)\int_{x_0}^{x}\frac{y_1(t)g(t)dt}{W(y_1,y_2)(t)}$$

$$= \int_{x_0}^{x}\frac{-y_1(x)y_2(t)g(t)dt}{W(y_1,y_2)(t)} + \int_{x_0}^{x}\frac{y_2(x)y_1(t)g(t)dt}{W(y_1,y_2)(t)}$$

$$= \int_{x_0}^{x}\frac{[y_1(t)y_2(x) - y_1(x)y_2(t)]g(t)dt}{y_1(t)y_2'(t) - y_1'(t)y_2(t)}.$$

25. Note that $y_1 = e^{\lambda x}\cos\mu x$ and $y_2 = e^{\lambda x}\sin\mu x$ and thus
$W(y_1,y_2) = \mu e^{2\lambda x}$. From Problem 22 we then have:

$$Y(x) = \int_{x_0}^{x}\frac{e^{\lambda t}\cos\mu t e^{\lambda x}\sin\mu x - e^{\lambda x}\cos\mu x e^{\lambda t}\sin\mu t}{\mu e^{2\lambda t}}g(t)dt$$

$$= \mu^{-1}\int_{x_0}^{x}e^{\lambda(x-t)}[\cos\mu t\sin\mu x - \cos\mu x\sin\mu t]g(t)dt$$

$$= \mu^{-1}\int_{x_0}^{x}e^{\lambda(x-t)}[\sin\mu(x-t)]g(t)dt.$$

29. First, we put the D.E. in standard form by dividing by
x^2: $y'' - 2y'/x + 2y/x^2 = 4$. Assuming that $y = xv(x)$ and
substituting in the D.E. we obtain $xv'' = 4$. Hence
$v'(x) = 4\ln x + c_2$ and

$v(x) = 4 \int \ln x \, dx + c_2 x = 4(x \ln x - x) + c_2 x$. The general
solution is $c_1 y_1(x) + x v(x) = c_1 x + 4(x^2 \ln x - x^2) + c_2 x^2$.
Since $-4x^2$ is a multiple of $y_2 = c_2 x^2$ we can write
$y = c_1 x + c_2 x^2 + 4x^2 \ln x$.

Section 3.8, Page 181

2. From Eq.(15) we have $R\cos\delta = -1$, and $R\sin\delta = \sqrt{13}$. Thus
    $R = \sqrt{1+3} = 2$ and $\delta = \tan^{-1}(-\sqrt{3}) + \pi = 2\pi/3 \cong 2.09440$.
    Note that we have to "add" π to the inverse tangent value
    since δ must be a second quadrant angle. Thus
    $u = 2\cos(t-2\pi/3)$.

6. The motion is an undamped free vibration. The units are
    in the CGS system. The spring constant
    $k = (100 \text{ gm})(980 \text{cm/sec}^2)/5\text{cm}$. Hence the D.E. for the
    motion is $100u'' + [(100 \cdot 980)/5]u = 0$ where u is measured
    in cm and time in sec. We obtain $u'' + 196u = 0$ so
    $u = A\cos 14t + B\sin 14t$. The I.C. are $u(0) = 0 \rightarrow A = 0$
    and $u'(0) = 10$ cm/sec $\rightarrow B = 10/14 = 5/7$. Hence
    $u(t) = (5/7)\sin 4t$, which first reaches equilibrium when
    $14t = \pi$, or $t = \pi/14$.

8. We use Eq.(31) without R (there is no resistor) and E(t)
    and with $L = 1$ henry and $1/C = 4 \times 10^6$ since $C = .25 \times 10^{-6}$
    farads. Thus the I.V.P. is $Q'' + 4 \times 10^6 Q = 0$, $Q(0) = 10^{-6}$
    coulombs and $Q'(0) = 0$.

9. The spring constant is $k = (20)(980)/5 = 3920$ dyne/cm.
    The I.V.P. for the motion is $20u'' + 400u' + 3920u = 0$ or
    $u'' + 20u' + 196u = 0$ and $u(0) = 2$, $u'(0) = 0$. Here u is
    measured in cm and t in sec. The general solution of the
    D.E. is $u = Ae^{-10t}\cos 4\sqrt{6}\,t + Be^{-10t}\sin 4\sqrt{6}\,t$. The I.C.
    $u(0) = 2 \rightarrow A = 2$ and $u'(0) = 0 \rightarrow -10A + 4\sqrt{6}\,B = 0$. The
    solution is $u = e^{-10t}[2\cos 4\sqrt{6}\,t + 5(\sin 4\sqrt{6}\,t)/\sqrt{6}\,]$ cm.
    The quasi frequency is $\mu = 4\sqrt{2}$, the quasi period is
    $T_d = 2\pi\mu = \pi/2\sqrt{6}$ and $T_d/T = 7/2\sqrt{6}$ since
    $T = 2\pi/14 = \pi/7$.

12. Substituting the given values for L, C and R we obtain
    the D.E. $.2Q'' + 3 \times 10^2 Q' + 10^5 Q = 0$. The I.C. are
    $Q(0) = 10^{-6}$ and $Q'(0) = I(0) = 0$. Assuming $Q = e^{rt}$, we
    obtain the roots of the characteristic equation as
    $r_1 = -500$ and $r_2 = -1000$. Thus $Q = c_1 e^{-500t} + c_2 e^{-1000t}$
    and hence $Q(0) = 10^{-6} \rightarrow c_1 + c_2 = 10^{-6}$ and

$Q'(0) = 0 \rightarrow -500c_1 - 1000c_2 = 0$. Solving for c_1 and c_2 yields the solution.

14. Note that $R\cos\delta = 0$ and $R\sin\delta = 1/8\sqrt{3}$. Since $\sin\delta > 0$, we conclude $\delta = \pi/2$ and $R = 1/8\sqrt{3}$. For $|u| \leq R/100$ we must have $e^{-2\tau} \leq 1/100$, or $\tau \geq (\ln 100)/2 \cong 2.3026$ sec.

19. The mass is $8/32$ lb-sec^2/ft, and the spring constant is $8/(1/8) = 64$ lb/ft. Hence $(1/4)u'' + \gamma u' + 64u = 0$ or $u'' + 4\gamma u' + 256u = 0$, where u is measured in ft, t in sec and the units of γ are lb-sec/ft. We look for solutions of the D.E. of the form $u = e^{rt}$ and find $r^2 + 4\gamma r + 256 = 0$, so $r_1, r_2 = [-4\gamma \pm \sqrt{16\gamma^2 - 1024}]/2$. The system will be overdamped, critically damped or underdamped as $(16\gamma^2 - 1024)$ is > 0, $= 0$, or < 0, respectively. Thus the system is critically damped when $\gamma = 8$ lb-sec/ft.

21. The general solution of the D.E. is $u = Ae^{r_1 t} + Be^{r_2 t}$ where $r_1, r_2 = [-\gamma \pm (\gamma^2 - 4km)^{1/2}]/2m$ provided $\gamma^2 - 4km \neq 0$, and where A and B are determined by the I.C. When the motion is overdamped, $\gamma^2 - 4km > 0$ and $r_1 > r_2$. Setting $u = 0$, we obtain $Ae^{r_1 t} = -Be^{r_2 t}$ or $e^{(r_1 - r_2)} = -B/A$. Since the exponential function is a monotone function, there is at most one value of t (when $B/A < 0$) for which this equation can be satisfied. Hence u can vanish at most once. If the system is critically damped, the general solution is $u(t) = (A + Bt)e^{-\gamma t/2m}$. The exponential function is never zero; hence u can vanish only if $A + Bt = 0$. If $B = 0$ then u never vanishes; if $B \neq 0$ then u vanishes once at $t = -A/B$ provided $A/B < 0$.

22. The general solution of Eq.(21) for the case of critical damping is $u = (A + Bt)e^{-\gamma t/2m}$. The I.C. $u(0) = u_0 \rightarrow A = u_0$ and $u'(0) = u_0' \rightarrow u_0' = A(-\gamma/2m) + B$. Hence $u = [u_0 + (u_0' + \gamma u_0/2m)t]e^{-\gamma t/2m}$. We now ask what conditions will insure that $u = 0$ at least once. Since the exponential function is never zero we require $u_0 + (u_0' + \gamma u_0/2m)t = 0$ at a positive value of t. This requires that $u_0' + \gamma u_0/2m \neq 0$ and that $t = -u_0(u_0' + \gamma u_0/2m)^{-1} > 0$. We know that $u_0 > 0$ so we must have $u_0' + \gamma u_0/2m < 0$ or $u_0' < -\gamma u_0/2m$.

28. First, consider the static case. Let $\Delta\ell$
     denote the length of the block below
     the surface of the water. The weight
     of the block, which is a downward force,
     is $w = \rho\ell^3 g$. This is balanced by an equal
     and opposite buoyancy force B, which is
     equal to the weight of the displaced
     water. Thus $B = (\rho_0\ell^2\Delta\ell)g = \rho\ell^3 g$ so $\rho_0\Delta\ell = \rho\ell$. Now let x
     be the displacement of the block from its equilibrium
     position. We take downward as the positive direction.
     In a displaced position the forces acting on the block
     are its weight, which acts downward and is unchanged, and
     the buoyancy force which is now $\rho_0\ell^2(\Delta\ell + x)g$ and acts
     upward. The resultant force must be equal to the mass of
     the block times the acceleration, namely $\rho\ell^3 x''$. Hence
     $\rho\ell^3 g - \rho_0\ell^2(\Delta\ell + x)g = \rho\ell^3 x''$. The D.E. for the motion of
     the block is $\rho\ell^3 x'' + \rho_0\ell^2 gx = 0$. This gives a simple
     harmonic motion with frequency $(\rho_0 g/\rho\ell)^{1/2}$ and natural
     period $2\pi(\rho\ell/\rho_0 g)^{1/2}$.

Section 3.9, Page 188

1.   We use the trigonometric identities
     $\cos(A \pm B) = \cos A \cos B \pm \sin A \sin B$ to obtain a formula
     for the difference of two cosines in terms of sines. We
     have $\cos(A + B) - \cos(A - B) = -2\sin A \sin B$. If we choose
     $A + B = 9t$ and $A - B = 7t$, then $A = 8t$ and $B = t$.
     Substituting in the formula just derived, we obtain
     $\cos 9t - \cos 7t = -2\sin 8t \sin t$.

5.   The mass $m = 4/32 = 1/8$ lb-sec^2/ft and the spring
     constant $k = 4/(1/8) = 32$ lb/ft. Since there is no
     damping, the I.V.P. is $(1/8)u'' + 32u = 2\cos 3t$,
     $u(0) = 1/6$, $u'(0) = 0$ where u is measured in ft and t in
     sec.

7a. From the solution to Problem 5, we have $m = 1/8$, $F_0 = 2$,
     $\omega_0^2 = 256$, and $\omega^2 = 9$, so Eq.(3) becomes
     $u = c_1\cos 16t + c_2\sin 16t + \dfrac{16}{247}\cos 3t$. The I.C.
     $u(0) = 1/6 \rightarrow c_1 + 16/247 = 1/6$ and $u'(0) = 0 \rightarrow 16c_2 = 0$,
     so the solution is $u = (151/1482)\cos 16t + (16/247)\cos 3t$
     ft.

7b. Resonance occurs when the frequency ω of the forcing
    function $4\sin\omega t$ is the same as the natural frequency ω_0
    of the system. Since $\omega_0 = 16$, the system will resonate
    when $\omega = 16$.

10. Note that this problem involves resonance and thus the
    particular solution has the form $t(A\cos 8t + B\sin 8t)$.

11a. For this problem the mass $m = 8/32$ lb-sec^2/ft and the
    spring constant $k = 8/(1/2) = 16$ lb/ft, so the D.E. is
    $0.25u'' + 0.25u' + 16u = 4\cos 2t$ where u is measured in ft
    and t in sec. To determine the steady state response we
    need only compute a particular solution of the
    nonhomogeneous D.E. since the solutions of the
    homogeneous D.E. decay to zero as $t \to \infty$. We assume
    $u(t) = A\cos 2t + B\sin 2t$, and substitute in the D.E.:
    $-A\cos 2t - B\sin 2t + (1/2)(-A\sin 2t + B\cos 2t) + 16(A\cos 2t +$
    $B\sin 2t) = 4\cos 2t$. Hence $15A + (1/2)B = 4$ and
    $-(1/2)A + 15B = 0$, from which we obtain $A = 240/901$ and
    $B = 8/901$. The steady state response is
    $u(t) = (240\cos 2t + 8\sin 2t)/901$.

11b. In order to determine the value of m that maximizes the
    steady state response, we note that the present problem
    has exactly the form of the problem considered in the
    text, Eqs.(8) and (9). The response is a maximum when
    $f(m) = m^2(\omega_0^2 - \omega^2)^2 + \gamma^2\omega^2$, where $\omega_0^2 = k/m$, is a
    minimum. We calculate df/dm, and set this quantity equal
    to zero, and obtain $m = k/\omega^2$. We verify that this value
    of m gives a minimum of f(m) by the second derivative
    test. For this problem $k = 16$ lb/ft and $\omega = 2$ rad/sec
    so the value of m that maximizes the response of the
    system is $m = 4$ slugs.

17. We must solve the three I.V.P.:   (1) $u_1'' + u_1 = F_0 t/m$,
    $0 < t < \pi$, $u_1(0) = u_1'(0) = 0$; (2) $u_2'' + u_2 = F_0(2\pi-t)/m$,
    $\pi < t < 2\pi$, $u_2(\pi) = u_1(\pi)$, $u_2'(\pi) = u_1'(\pi)$; and
    (3) $u_3'' + u_3 = 0$, $2\pi < t$, $u_3(2\pi) = u_2(2\pi)$, $u_3'(2\pi) = u_2'(2\pi)$.
    The conditions at π and 2π insure the continuity of u and
    u' at those points. The general solutions of the D.E.
    are $u_1 = b_1\cos t + b_2\sin t + F_0 t/m$, $u_2 = c_1\cos t + c_2\sin t +$
    $F_0(2\pi-t)/m$, and $u_3 = d_1\cos t + d_2\sin t$. The I.C.and
    matching conditions, in order, give $b_1 = 0$, $b_2 + F_0/m = 0$,
    $-b_1 + \pi F_0/m = -c_1 + \pi F_0/m$, $-b_2 + F_0/m = -c_2 - F_0/m$, $c_1 = d_1$,

and $c_2 - F_0/m = d_2$. Solving these equations we obtain

$$u = (F_0/m) \begin{cases} t - \sin t & , \ 0 \leq t \leq \pi \\ (2\pi - t) - 3\sin t & , \ \pi < t \leq 2\pi \\ - 4\sin t & , \ 2\pi < t. \end{cases}$$

18. The I.V.P. is $Q'' + 5 \times 10^3 Q' + 4 \times 10^6 Q = 12$, $Q(0) = 0$, and
$Q'(0) = 0$. The particular solution is of the form $Q = A$,
so that upon substitution into the D.E. we obtain
$4 \times 10^6 A = 12$ or $A = 3 \times 10^{-6}$. The general solution of the
D.E. is $Q = c_1 e^{r_1 t} + c_2 e^{r_2 t} + 3 \times 10^{-6}$, where r_1 and r_2 satisfy
$r^2 + 5 \times 10^3 r + 4 \times 10^6 = 0$ and thus are $r_1 = -1000$ and
$r_2 = -4000$. The I.C. yield $c_1 = -4 \times 10^{-6}$ and $c_2 = 10^{-6}$ and
thus $Q = 10^{-6}(e^{-4000t} - 4e^{-1000t} + 3)$ coulombs. Substituting
$t = .001$ sec we obtain
$Q(.001) = 10^{-6}(e^{-4} - 4e^{-1} + 3) = 10^{-6}(0.0183 - 4 \times 0.3679 +$
$3) = 1.55 \times 10^{-6}$ coulombs. Since exponentials are to a
negative power $Q(t) \to 3 \times 10^{-6}$ coulombs as $t \to \infty$, and
this is the steady state charge.

22. The amplitude of the steady state response is seven or
eight times the amplitude (3) of the forcing term. This
large an increase is due to the fact that the forcing
function has the same frequency as the natural frequency
of the system.
There also appears to be a phase lag of approximately 1/4
of a period. That is, the maximum of the response occurs
1/4 of a period after the maximum of the forcing
function. Both these results are substantially different
than those of either Problems 21 or 23.

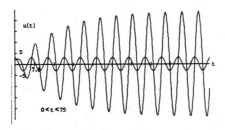

CHAPTER 4

Section 4.1, Page 194

2. Writing the equation in standard form, we obtain
    $y''' + (\sin x/x)y'' + (3/x)y = \cos x/x$. The functions
    $p_1(x) = \sin x/x$, $p_3(x) = 3/x$ and $g(x) = \cos x/x$ have
    discontinuities at $x = 0$. Hence Theorem 4.1.1 guarantees
    that a solution exists for $x < 0$ and for $x > 0$.

10. Differentiating $y(x)$ as given we obtain
    $y' = c_1 + 2c_2 x + 3c_3 x^2$, $y'' = 2c_2 + 6c_3 x$ and $y''' = 6c_3$. The
    last equation gives $c_3 = y'''/6$ which when substituted into
    the equation for y yields $c_2 = (y'' - xy''')/2$. Substituting
    for c_2 and c_3 into the equation for y' and solving for c_1
    results in $c_1 = y' - xy'' + x^2 y'''/2$. Finally inserting
    these values of c_1, c_2 and c_3 into the equation for $y(x)$
    and simplifying yields the desired result.

15. That e^x, e^{-x}, and e^{-2x} are solutions can be verified by
    direct substitution. Computing the Wronskian we obtain,

    $$W(e^x, e^{-x}, e^{-2x}) = \begin{vmatrix} e^x & e^{-x} & e^{-2x} \\ e^x & -e^{-x} & -2e^{-2x} \\ e^x & e^{-x} & 4e^{-2x} \end{vmatrix} = -6e^{-2x}.$$

19. To show that the given Wronskian is zero, it is helpful
    to note that $(\sin^2 x)' = 2\sin x\cos x = \sin 2x$. This result
    can be obtained directly since $\sin^2 x = (1 - \cos 2x)/2 = \dfrac{1}{10}(5) + (-1/2)\cos 2x$ and hence $\sin^2 x$ is a linear
    combination of 5 and $\cos 2x$. Thus the functions are
    linearly dependent and their Wronskian is zero.

21c. If we let $L[y] = y^{iv} - 5y'' + 4y$ and if we use the result
    of Problem 6b, we have $L[e^{rx}] = (r^4 - 5r^2 + 4)e^{rx}$. Thus e^{rx}
    will be a solution of the D.E. provided $(r^2-4)(r^2-1) = 0$.
    Solving for r, we obtain the four soltions e^x, e^{-x}, e^{2x}
    and e^{-2x}. Since $W(e^x, e^{-x}, e^{2x}, e^{-2x}) \neq 0$, the four
    functions form a fundamental set of solutions.

25. As in Problem 24, let $y = v(x)e^x$. Differentiating three

times and substituting into the D.E. yields
$(2-x)e^x v''' + (3-x)e^x v'' = 0$. Dividing by $(2-x)e^x$ and
letting $w = v''$ we obtain the first order separable

equation $w' = -\dfrac{x-3}{x-2}w = (-1 + \dfrac{1}{x-2})w$. Separating x and w,

integrating, and then solving for w yields
$w = v'' = c_1(x-2)e^{-x}$. Integrating this twice then gives
$v = c_1 x e^{-x} + c_2 x + c_3$ so that $y = v e^x = c_1 x + c_2 x e^x + c_3 e^x$,
which is the complete solution, since it contains the
given $y_1(x)$ and three constants.

Section 4.2, Page 200

2.   If $-1 + i\sqrt{3} = Re^{i\theta}$, then $R = [(-1)^2 + (\sqrt{3})^2]^{1/2} = 2$. The
     angle θ is given by $R\cos\theta = 2\cos\theta = -1$ and
     $R\sin\theta = 2\sin\theta = \sqrt{3}$. Hence $\cos\theta = -1/2$ and
     $\sin\theta = \sqrt{3}/2$ which has the solution $\theta = 2\pi/3$. The angle
     θ is only determined up to an additive integer multiple
     of $\pm 2\pi$.

8.   Writing $(1-i)$ in the form $Re^{i\theta}$, we obtain
     $(1-i) = \sqrt{2}\,e^{i(-\pi/4+2m\pi)}$ where m is any integer. Hence,
     $(1-i)^{1/2} = [2^{1/2}e^{i(-\pi/4+2m\pi)}]^{1/2} = 2^{1/4}e^{i(-\pi/8+m\pi)}$. We obtain the
     two square roots by setting $m = 0,1$. They are
     $2^{1/4}e^{-i\pi/8}$ and $2^{1/4}e^{i3\pi/8}$. Note that any other integer value
     of m gives one of these two values. Also note that $1-i$
     could be written as $1-i = \sqrt{2}\,e^{i(7\pi/4 + 2m\pi)}$.

12. We look for solutions of the form $y = e^{rx}$. Substituting
     in the D.E., we obtain the characteristic equation
     $r^3 - 3r^2 + 3r - 1 = 0$ which has roots $r = 1,1,1$. Since
     the roots are repeated, the general solution is
     $y = c_1 e^x + c_2 x e^x + c_3 x^2 e^x$.

15. We look for solutions of the form $y = e^{rx}$. Substituting
     in the D.E. we obtain the characteristic equation
     $r^6 + 1 = 0$. The six roots of -1 are obtained by setting
     $m = 0,1,2,3,4,5$ in $(-1)^{1/6} = e^{i(\pi+2m\pi)/6}$. They are
     $e^{i\pi/6} = (\sqrt{3} + i)/2$, $e^{i\pi/2} = i$, $e^{i5\pi/6} = (-\sqrt{3} + i)/2$,
     $e^{i7\pi/6} = (-\sqrt{3} - i)/2$, $e^{i3\pi/2} = -i$, and
     $e^{i11\pi/6} = (\sqrt{3} - i)/2$. Note that there are three pairs of

conjugate roots. The general solution is
$$y = e^{\sqrt{3}\,x/2}[c_1\cos(x/2) + c_2\sin(x/2)] + e^{-\sqrt{3}\,x/2}[c_3\cos(x/2) + c_4\sin(x/2)] + c_5\cos x + c_6\sin x.$$

23. The characteristic equation is $r^3 + r = 0$ and hence
    $r = 0, +i, -i$ are the roots and the general solution is
    $y(x) = c_1 + c_2\cos x + c_3\sin x$. $y(0) = 0$ implies
    $c_1 + c_2 = 0$, $y'(0) = 1$ implies $c_3 = 1$ and $y''(0) = 2$
    implies $-c_2 = 2$. Use this last equation in the first to
    find $c_1 = 2$ and thus $y(x) = 2 - 2\cos x + \sin x$.

25. The general solution would normally be written
    $y(x) = c_1 + c_2 x + c_3 e^{2x} + c_4 x e^{2x}$. However, in order to
    evaluate the c's when the initial conditions are given at
    $x = 1$, it is advantageous to rewrite
    $y(x)$ as $y(x) = c_1 + c_2 x + c_5 e^{2(x-1)} + c_6(x-1)e^{2(x-1)}$.

27. The approach developed in this section for solving the
    D.E. would normally yield $y(x) = c_1\cos x + c_2\sin x + c_5 e^x +
    c_6 e^{-x}$ as the solution. Now use the definition of $\cosh x$
    and $\sinh x$ to yield the desired result. It is convenient
    to use $\cosh x$ and $\sinh x$ rather than e^x and e^{-x} because the
    I.C. are given at $x = 0$. Since $\cosh x$ and $\sinh x$ and all
    of their derivatives are either 0 or 1 at $x = 0$, the
    algebra in satisfying the I.C. is simplified.

28a. As in Section 3.8, the force that the spring designated
     by k_1 exerts on mass m_1 is $-3u_1$. By an analysis similar
     to that shown in Section 3.8, the middle spring exerts a
     force of $-2(u_1-u_2)$ on mass m_1 and a force of $-2(u_2-u_1)$ on
     mass m_2. In all cases the positive direction is taken in
     the direction shown in Figure 4.2.1.

28c. From Eq.(i) we have $u_1''(0) = 2u_2(0) - 5u_1(0) = -1$ and
     $u_1'''(0) = 2u_2'(0) - 5u_1'(0) = 0$. From Prob.28b we have
     $u_1 = c_1\cos t + c_2\sin t + c_3\cos\sqrt{6}\,t + c_4\sin\sqrt{6}\,t$. Thus
     $c_1+c_3 = 1$, $c_2+\sqrt{6}\,c_4 = 0$, $-c_1-6c_3 = -1$ and $-c_2-6\sqrt{6}\,c_4 = 0$,
     which yield $c_1 = 1$ and $c_2 = c_3 = c_4 = 0$, so that
     $u_1 = \cos t$. The first of Eqs.(i) then gives u_2.

29. Since $a_0 = 1 > 0$, the quantities that must be checked are

$$3, \quad \begin{vmatrix} 3 & 1 \\ 1 & 3 \end{vmatrix}, \text{ and } \begin{vmatrix} 3 & 1 & 0 \\ 1 & 3 & 0 \\ 0 & 0 & 1 \end{vmatrix}.$$

They are each positive; hence the D.E. is asympotically stable. In this case the characteristic equation is $r^3 + 3r^2 + 3r + 1 = (r+1)^3 = 0$, which has the roots $r_1 = r_2 = r_3 = -1$. Thus $y = (c_1 + c_2 + c_3 x^2)e^{-x}$, which decays to zero as $x \to \infty$ using L'Hopital's Rule.

33. Again $a_0 > 0$ and the quantities that must be checked are

$$0.1, \quad \begin{vmatrix} 0.1 & 1 \\ -0.4 & 1.2 \end{vmatrix}, \text{ and } \begin{vmatrix} 0.1 & 1 & 0 \\ -0.4 & 1.2 & .1 \\ 0 & 0 & -0.4 \end{vmatrix}.$$

The last quantity has the value $-0.4(0.12 + 0.4) < 0$. Hence the D.E. is unstable.

Section 4.3, Page 205

1. First solve the homogeneous D.E. The characteristic equation is $r^3 - r^2 - r + 1 = 0$, and the roots are $r = -1$, 1, 1; hence $y_c(x) = c_1 e^{-x} + c_2 e^x + c_3 x e^x$. Using the superposition principle, we can write a particular solution as the sum of particular solutions corresponding to the D.E. $y''' - y'' - y' + y = 2e^{-x}$ and $y''' - y'' - y' + y = 3$. Our initial choice for a particular solution, Y_1, of the first equation is Ae^{-x}; but e^{-x} is a solution of the homogeneous equation so we multiply by x. Thus, $Y_1(x) = Axe^{-x}$. For the second equation we choose $Y_2(x) = B$, and there is no need to modify this choice. The constants are determined by substituting into the individual equations. We obtain $A = 1/2$, $B = 3$. Thus, the general solution is
$$y = c_1 e^{-x} + c_2 e^x + c_3 x e^x + 3 + (xe^{-x})/2.$$

9. The characteristic equation for the related homogeneous D.E. is $r^3 + 4r = 0$ with roots $r = 0$, $+2i$, $-2i$. Hence $y_c(x) = c_1 + c_2\cos 2x + c_3\sin 2x$. The initial choice for $Y(x)$ is $Ax + B$, but since B is a solution of the homogeneous equation we must multiply by x and assume $Y(x) = x(Ax+B)$. A and B are found by substituting in the

D.E., which gives $A = 1/8$, $B = 0$, and thus the general solution is $y(x) = c_1 + c_2\cos 2x + c_3\sin 2x + (1/8)x^2$. Applying the I.C. we have $y(0) = 0 \rightarrow c_1 + c_2 = 0$, $y'(0) = 0 \rightarrow 2c_3 = 0$, and $y''(0) = 1 \rightarrow -4c_2 + 1/4 = 1$, which have the solution $c_1 = 3/16$, $c_2 = -3/16$, $c_3 = 0$.

12. The characteristic equation for the homogeneous D.E. is $r^3 - 2r^2 + r = 0$ with roots $r = 0, 1, 1$. Hence the complementary solution is $y_c(x) = c_1 + c_2 e^x + c_3 x e^x$. We consider the differential equations $y''' - 2y'' + y' = x^3$ and $y''' - 2y'' + y' = 2e^x$ separately. Our initial choice for a particular solution, Y_1, of the first equation is $A_0 x^3 + A_1 x^2 + A_2 x + A_3$; but since a constant is a solution of the homogeneous equation we must multiply by x. Thus $Y_1(x) = x(A_0 x^3 + A_1 x^2 + A_2 x + A_3)$. For the second equation we first choose $Y_2(x) = Be^x$, but since both e^x and xe^x are solutions of the homogeneous equation, we multiply by x^2 to obtain $Y_2(x) = Bx^2 e^x$. Then $Y(x) = Y_1(x) + Y_2(x)$ by the superposition principle and $y(x) = y_c(x) + Y(x)$.

16. The complementary solution is $y_c(x) = c_1 + c_2 e^{-x} + c_3 e^x + c_4 x e^x$. The superposition principle allows us to consider separately the D.E. $y^{iv} - y''' - y'' + y' = x^2 + 4$ and $y^{iv} - y''' - y'' + y' = x\sin x$. For the first equation our initial choice is $Y_1(x) = A_0 x^2 + A_1 x + A_2$; but this must be multiplied by x since a constant is a solution of the homogeneous D.E. Hence $Y_1(x) = x(A_0 x^2 + A_1 x + A_2)$. For the second equation our initial choice that $Y_2 = (B_0 x + B_1)\cos x + + (C_0 x + C_1)\sin x$ does not need to be modified. Hence $Y(x) = x(A_0 x^2 + A_1 x + A_2) + (B_0 x + B_1)\cos x + (C_0 x + C_1)\sin x$.

19. $(D-a)(D-b)f = (D-a)(Df-bf) = D^2 f - (a+b)Df + abf$ and $(D-b)(D-a)f = (D-b)(Df-af) = D^2 f - (b+a)Df + baf$. Since $a+b = b+a$ and $ab = ba$, we find the given equation holds for any function f.

21a. The D.E. of Problem 12 can be written as $D(D-1)^2 y = x^3 + 2e^x$. Since D^4 annihilates x^3 and $(D-1)$

annihilates $2e^x$, we have $D^5(D-1)^3y = 0$, which corresponds
to Eq.(ii) of Problem 20. The solution of this equation
is $y(x) = A_1x^4 + A_2x^3 + A_3x^2 + A_4x + A_5 +$
$(B_1x^2 + B_2x + B_3)e^{-x}$. Since $A_5 + (B_2x + B_3)e^{-x}$ are
solutions of the homogeneous equation related to the
original D.E., they may be deleted and thus
$Y(x) = A_1x^4 + A_2x^3 + A_3x^2 + A_4x + B_1x^2e^{-x}$.

21b. $(D+1)^2(D^2+1)$ annihilates the right side of the D.E. of
Problem 13.

21e. $D^3(D^2+1)^2$ annihilates the right side of the D.E. of
Problem 16.

Section 4.4, Page 210

1.   The complementary solution is $y_c = c_1 + c_2\cos x + c_3\sin x$
and thus we assume a particular solution of the form
$Y = u_1(x) + u_2(x)\cos x + u_3(x)\sin x$. Differentiating and
assuming Eq.(5), we obtain $Y' = -u_2\sin x + u_3\cos x$ and
$$u_1' + u_2'\cos x + u_3'\sin x = 0 \qquad (a).$$
Continuing this process we obtain $Y'' = -u_2\cos x - u_3\sin x$,
$Y''' = u_2\sin x - u_3\cos x - u_2'\cos x - u_3'\sin x$ and
$$-u_2'\sin x + u_3'\cos x = 0 \qquad\qquad (b).$$
Substituting Y and its derivatives, as given above, into
the D.E. we obtain the third equation:
$$-u_2'\cos x - u_3'\sin x = \tan x \qquad (c).$$
Equations (a), (b) and (c) constitute Eqs.(10) of the
text for this problem and may be solved to give
$u_1' = \tan x$, $u_2' = -\sin x$, and $u_3' = -\sin^2 x/\cos x$.

5.   Since a fundamental set of solutions of the homogeneous
D.E. is $y_1 = e^x$, $y_2 = \cos x$, $y_3 = \sin x$, a particular
solution is of the form $Y(x) = e^xu_1(x) + (\cos x)u_2(x) +$
$(\sin x)u_3(x)$. Differentiating and making the same
assumptions that lead to Eqs.(10), we obtain
$$u_1'e^x + u_2'\cos x + u_3'\sin x = 0$$
$$u_1'e^x - u_2'\sin x + u_3'\cos x = 0$$
$$u_1'e^x - u_2'\cos x - u_3'\sin x = g(x)$$
Solving these equations using either determinants or by

elimination, we obtain $u_1' = (1/2)e^{-x}g(x)$,

$u_2' = (1/2)(\sin x - \cos x)g(x)$, $u_3' = -(1/2)(\sin x + \cos x)g(x)$.

Integrating these and substituting into Y yields

$$Y(x) = \frac{1}{2}\{e^x\int_{x_0}^x e^{-t}g(t)dt + \cos x\int_{x_0}^x(\sin t - \cos t)g(t)dt$$

$$-\sin x\int_{x_0}^x(\sin t + \cos t)g(t)dt\}.$$

This can be written in the form

$$Y(x) = (1/2)\int_{x_0}^x(e^{x-t} + \cos x\sin t - \cos x\cos t$$

$$-\sin x\sin t - \sin x\cos t)g(t)dt.$$

If we use the trigonometric identities $\sin(A-B) = \sin A\cos B - \cos A\sin B$ and $\cos(A-B) = \cos A\cos B + \sin A\sin B$, we obtain the desired result. Note: Eqs.(11) and (12) of this section give the same result, but it is not recommended to memorize these equations.

7.   The particular solution has the form $Y = e^x u_1(x) + xe^x u_2(x) + x^2 e^x u_3(x)$. Differentiating, making the same assumptions as in the earlier problems, and solving the three linear equations for u_1', u_2', and u_3' yields

$u_1' = (1/2)x^2 e^{-x}g(x)$, $u_2' = -xe^{-x}g(x)$ and $u_3' = (1/2)e^{-x}g(x)$.

Integrating and substituting into Y yields the desired solution. For instance

$$xe^x u_2 = -xe^x\int_{x_0}^x te^{-t}g(t)dt = -\frac{1}{2}\int_{x_0}^x 2xte^{(x-t)}g(t)dt,\text{ and}$$

likewise for u_1 and u_3. If $g(x) = x^{-2}e^x$ then $g(t) = e^t/t^2$ and the integration is accomplished using the power rule. Note that terms involving x_0 become part of the complimentary solution.

CHAPTER 5

2. Use the ratio test:
$$\lim_{n \to \infty} \frac{\left| (n+1) x^{n+1}/2^{n+1} \right|}{\left| nx^n/2^n \right|} = \lim_{n \to \infty} \frac{n+1}{n} \frac{1}{2} |x| = \frac{|x|}{2}.$$
Therefore the series converges absolutely for $|x| < 2$. For $x = 2$ and $x = -2$ the nth term does not approach zero as $n \to \infty$ so the series diverge. Hence the radius of convergence is $\rho = 2$.

5. Use the ratio test:
$$\lim_{n \to \infty} \frac{\left| (2x+1)^{n+1}/(n+1)^2 \right|}{(2x+1)^n/n^2} = \lim_{n \to \infty} \frac{n^2}{(n+1)^2} |2x+1| = |2x+1|.$$
Therefore the series converges absolutely for $|x+1/2| < 1/2$. At $x = 0$ and $x = -1$ the series also converge absolutely. However, for $|x+1/2| > 1/2$ the series diverges by the ratio test. The radius of convergence is $\rho = 1/2$.

9. For this problem $f(x) = \sin x$. Hence $f'(x) = \cos x$, $f''(x) = -\sin x$, $f'''(x) = -\cos x, \ldots$. Then $f(0) = 0$, $f'(0) = 1$, $f''(0) = 0$, $f'''(0) = -1,\ldots$. The even terms in the series will vanish and the odd terms will alternate in sign. We obtain $\sin x = \sum_{n=0}^{\infty} (-1)^n x^{2n+1}/(2n+1)!$. From the ratio test it follows that $\rho = \infty$.

12. For this problem $f(x) = x^2$. Hence $f'(x) = 2x$, $f''(x) = 2$, and $f^{(n)}(x) = 0$ for $n > 2$. Then $f(-1) = 1$, $f'(-1) = -2$, $f''(-1) = 2$ and $x^2 = 1 - 2(x+1) + 2(x+1)^2/2! = 1 - 2(x+1) + (x+1)^2$. Since the series terminates after a finite number of terms, it converges for all x. Thus $\rho = \infty$.

13. For this problem $f(x) = \ln x$. Hence $f'(x) = 1/x$, $f''(x) = -1/x^2$, $f'''(x) = 1\cdot2/x^3,\ldots$, and $f^{(n)}(x) = (-1)^{n+1}(n-1)!/x^n$. Then $f(1) = 0$, $f'(1) = 1$, $f''(1) = -1$, $f'''(1) = 1\cdot2,\ldots$, $f^{(n)}(1) = (-1)^{n+1}(n-1)!$ The Taylor series is $\ln x = (x-1) - (x-1)^2/2 + (x-1)^3/3 - \ldots = \sum_{n=1}^{\infty} (-1)^{n+1}(x-1)^n/n$. It follows from the ratio test that

the series converges absolutely for $|x-1| < 1$. However, the series diverges at $x = 0$ so $\rho = 1$.

19. Set $m = n-1$ on the right hand side of the equation. Then $n = m+1$ and when $n = 1$, $m = 0$. Thus the right hand side

becomes $\sum\limits_{m=0}^{\infty} a_m(x-1)^{m+1}$, which is the same as the left hand

side when m is replaced by n.

24. If we shift the index of summation in the first sum by letting $m = n-1$, we have

$$\sum_{n=1}^{\infty} n\, a_n x^{n-1} = \sum_{m=0}^{\infty} (m+1)\, a_{m+1} x^m.$$ Substituting this into

the given equation and letting $m = n$ again, we obtain:

$$\sum_{n=0}^{\infty} (n+1) a_{n+1}\, x^n + 2 \sum_{n=0}^{\infty} a_n x^n = 0,\ \text{or}$$

$$\sum_{n=0}^{\infty} [(n+1) a_{n+1} + 2a_n] x^n = 0.$$

Hence $a_{n+1} = -2a_n/(n+1)$ for $n = 0,1,2,3,\ldots$. Thus $a_1 = -2a_0$, $a_2 = -2a_1/2 = 2^2 a_0/2$, $a_3 = -2a_2/3 = -2^3 a_0/2\cdot3 = -2^3 a_0/3!\ldots$ and $a_n = (-1)^n 2^n a_0/n!$. Notice that for $n = 0$ this formula reduces to a_0 so we can write

$$\sum_{n=0}^{\infty} a_n x^n = \sum_{n=0}^{\infty} (-1)^n 2^n\, a_0 x^n/n! = a_0 \sum_{n=0}^{\infty} (-2x)^n/n! = a_0 e^{-2x}.$$

Section 5.2, Page 226

2. $y = \sum\limits_{n=0}^{\infty} a_n x^n$; $y' = \sum\limits_{n=1}^{\infty} n a_n x^{n-1}$ and since we must multiply

y' by x in the D.E. we do not shift the index; and

$$y'' = \sum_{n=2}^{\infty} n(n-1) a_n x^{n-2} = \sum_{n=0}^{\infty} (n+2)(n+1) a_{n+2} x^n.$$ Substituting

in the D.E., we obtain

$$\sum_{n=0}^{\infty} (n+2)(n+1) a_{n+2} x^n - \sum_{n=1}^{\infty} n a_n x^n - \sum_{n=0}^{\infty} a_n x^n = 0.$$ In order to

have the starting point the same in all three summations,

we let $n = 0$ in the first and third terms to obtain the following

$$(2 \cdot 1 \, a_2 - a_0) x^0 + \sum_{n=1}^{\infty} [(n+2)(n+1) a_{n+2} - (n+1) a_n] x^n = 0.$$

Thus $a_{n+2} = a_n/(n+2)$ for $n = 1, 2, 3, \ldots$. However, note that recurrence relation is also correct for $n = 0$. The even and odd coefficients can be determined independently. We show how to calculate the odd a's:
$a_3 = a_1/3$, $a_5 = a_3/5 = a_1/5 \cdot 3$, $a_7 = a_5/7 = a_1/7 \cdot 5 \cdot 3, \ldots$.
Now notice that $a_3 = 2a_1/(2 \cdot 3) = 2a_1/3!$, that

$a_5 = 2 \cdot 4 a_1/(2 \cdot 3 \cdot 4 \cdot 5) = 2^2 2 a_1/5!$, and that

$a_7 = 2 \cdot 4 \cdot 6 a_1/(2 \cdot 3 \cdot 4 \cdot 5 \cdot 6 \cdot 7) = 2^3 3! \, a_1/7!$. Likewise

$a_9 = a_7/9 = 2^3 3! \, a_1/(7!)9 = 2^3 3! \, 8a_1/9! = 2^4 4! \, a_1/9!$.

Continuing we have $a_{2m+1} = 2^m m! \, a_1/(2m+1)!$. In the same way we find that the even a's are given by $a_{2m} = a_0/2^m m!$. Thus

$$y = a_0 \sum_{m=0}^{\infty} \frac{x^{2m}}{2^m m!} + a_1 \sum_{m=0}^{\infty} \frac{2^m m! \, x^{2m+1}}{(2m+1)!}.$$

3. $y = \sum_{n=0}^{\infty} a_n (x-1)^n$; $y' = \sum_{n=1}^{\infty} n a_n (x-1)^{n-1}$ and

$$y'' = \sum_{n=2}^{\infty} n(n-1) a_n (x-1)^{n-2} = \sum_{n=0}^{\infty} (n+2)(n+1) a_{n+2}(x-1)^n.$$

Substituting in the D.E. and replacing the coefficient x by $1 + (x-1)$ we obtain

$$\sum_{n=0}^{\infty} (n+2)(n+1) a_{n+2}(x-1)^n - \sum_{n=0}^{\infty} (n+1) a_{n+1}(x-1)^n - \sum_{n=1}^{\infty} n a_n (x-1)^n$$

$$- \sum_{n=0}^{\infty} a_n (x-1)^n = 0,$$

where the index has been shifted in the second summation. Letting $n = 0$ in the first, second, and the fourth sums, we obtain

$$(2 \cdot 1 \cdot a_2 - 1 \cdot a_1 - a_0)(x-1)^0 + \sum_{n=1}^{\infty} [(n+2)(n+1) a_{n+2}$$

$$- (n+1) a_{n+1} - (n+1) a_n] (x-1)^n = 0.$$

Thus $(n+2)a_{n+2} - a_{n+1} - a_n = 0$ for $n = 0,1,2,\ldots$. This
recurrance relation can be used to solve for a_2 in terms
of a_0 and a_1, then for a_3 in terms of a_0 and a_1, etc.   In
many cases it is easier to first take $a_0 = 0$ and generate
one solution and then take $a_1 = 0$ and generate the second
linearly independent solution. Thus, choosing $a_0 = 0$ we
find that $a_2 = a_1/2$, $a_3 = (a_2+a_1)/3 = a_1/2$,
$a_4 = (a_3+a_2)/4 = a_1/4$, $a_5 = (a_4+a_3)/5 = 3a_1/20,\ldots$. This
yields the solution $y_2(x) = a_1[(x-1) + (x-1)^2/2 +$
$(x-1)^3/2 + (x-1)^4/4 + 3(x-1)^5/20 + \ldots]$. The second
independent solution may be obtained by choosing $a_1 = 0$.
Then $a_2 = a_0/2$, $a_3 = (a_2+a_1)/3 = a_0/6$,
$a_4 = (a_3+a_2)/4 = a_0/6$, $a_5 = (a_4+a_3)/5 = a_0/15,\ldots$. This
yields the solution
$y_1(x) = a_0[1+(x-1)^2/2+(x-1)^3/6+(x-1)^4/6+(x-1)^5/15+\ldots]$.
Note that to obtain the answer given in the text we can
take $a_1 = 1$ in the y_2 solution and $a_0 = 1$ in the y_1
solution. The general solution is a linear combination
of y_1 and y_2.

5. $y = \sum_{n=0}^{\infty} a_n x^n$; $y' = \sum_{n=1}^{\infty} na_n x^{n-1}$; and $y'' = \sum_{n=2}^{\infty} 2n(n-1)a_n x^{n-2}$.

Substituting in the D.E. and shifting the index in the
series for y'' gives

$\sum_{n=0}^{\infty} (n+2)(n+1)a_{n+2}x^n - \sum_{n=1}^{\infty} (n+1)n\, a_{n+1}x^n + \sum_{n=0}^{\infty} a_n x^n = 0,$

$(2 \cdot 1 \cdot a_2 + a_0)x^0 + \sum_{n=1}^{\infty} [(n+2)(n+1)a_{n+2} - (n+1)na_{n+1} + a_n]x^n =$

0.   Thus $a_2 = -a_0/2$ and $a_{n+2} = na_{n+1}/(n+2) - a_n/(n+2)(n+1)$,
$n = 1,2,\ldots$. Choosing $a_0 = 0$ yields $a_2 = 0$, $a_3 = -a_1/6$,
$a_4 = 2a_3/4 = -a_1/12,\ldots$ which gives one solution as
$y_2(x) = a_1(x - x^3/6 - x^4/12 + \ldots)$. A linearly
independent solution is obtained by choosing $a_1 = 0$. Then
$a_2 = -a_0/2$, $a_3 = a_2/3 = -a_0/6$, $a_4 = 2a_3/4 - a_2/12 = -a_0/24,\ldots$
which gives $y_1(x) = a_0(1 - x^2/2 - x^3/6 - x^4/24 + \ldots)$.

14. You will need to rewrite x+1 as 3 + (x-2) in order to multiply x+1 times y' as a power series about $x_0 = 2$.

16. From Problem 6 we have
$$y(x) = c_1(1 - x^2 + \frac{1}{6}x^4 + \ldots) + c_2(x - \frac{1}{4}x^3 + \frac{7}{160}x^5 + \ldots).$$
Now $y(0) = c_1 = -1$ and $y'(0) = c_2 = 3$ and thus
$$y(x) = -1 + x^2 - \frac{1}{6}x^4 + 3x - \frac{3}{4}x^3 = -1 + 3x + x^2 - \frac{3}{4}x^3 - \ldots .$$

19. The D.E. transforms into $u''(t) + t^2 u'(t) + (t^2 + 2t)u(t) = 0$.

Assuming that $u(t) = \sum_{n=0}^{\infty} a_n t^n$, we have $u'(t) = \sum_{n=1}^{\infty} n a_n t^{n-1}$ and

$u''(t) = \sum_{n=2}^{\infty} n(n-1) a_n t^{n-2}$. Substituting in the D.E. and shifting indices yields

$$\sum_{n=0}^{\infty} (n+2)(n+1) a_{n+2} t^n + \sum_{n=2}^{\infty} (n-1) a_{n-1} t^n + \sum_{n=2}^{\infty} a_{n-2} t^n$$

$$+ \sum_{n=1}^{\infty} 2 a_{n-1} t^n = 0,$$

$$2 \cdot 1 \cdot a_2 t^0 + (3 \cdot 2 \cdot a_3 + 2 \cdot a_0) t^1 + \sum_{n=2}^{\infty} [(n+2)(n+1) a_{n+2}$$

$$+ (n+1) a_{n-1} + a_{n-2}] t^n = 0.$$

It follows that $a_2 = 0$, $a_3 = -a_0/3$ and
$a_{n+2} = -a_{n-1}/n+2 - a_{n-2}/[(n+2)(n+1)]$, $n = 2,3,4 \ldots$. We
obtain one solution by choosing $a_1 = 0$. Then $a_4 = -a_0/12$,
$a_5 = -a_2/5 - a_1/20 = 0$, $a_6 = -a_3/6 - a_2/30 = a_0/18, \ldots$. Thus
one solution is $u_1(t) = a_0(1 - t^3/3 - t^4/12 + t^6/18 + \ldots)$ so
$y_1(x) = u_1(x-1) = a_0[1 - (x-1)^3/3 - (x-1)^4/12 + (x-1)^6/18 + \ldots]$.
We obtain a second solution by choosing $a_0 = 0$. Then
$a_4 = -a_1/4$, $a_5 = -a_2/5 - a_1/20 = -a_1/20$,
$a_6 = -a_3/6 - a_2/30 = 0$, $a_7 = -a_4/7 - a_3/42 = a_1/28, \ldots$.
Thus a second linearly independent solution is

$u_2(t) = a_1[t - t^4/4 - t^5/20 + t^7/28 + ...]$ or

$y_2(x) = u_2(x-1)$

$\quad = a_1[(x-1) - (x-1)^4/4 - (x-1)^5/20 + (x-1)^7/28 + ...]$.

The Taylor series for $x^2 - 1$ about $x = 1$ may be obtained by writing $x = (x-1) + 1$ so $x^2 = (x-1)^2 + 2(x-1) + 1$ and $x^2 - 1 = (x-1)^2 + 2(x-1)$. The D.E. now appears as $y'' + (x-1)^2 y' + [(x-1)^2 + 2(x-1)]y = 0$ which is identical to the transformed equation with $t = x - 1$.

22. $y = a_0 + a_1 x + a_2 x^2 + ...$, $y^2 = a_0^2 + 2a_0 a_1 x + (2a_0 a_2 + a_1^2)x^2$
$+ ...$, $y' = a_1 + 2a_2 x + 3a_3 x^2 + ...$, and

$(y')^2 = a_1^2 + 4a_1 a_2 x + (6a_1 a_3 + 4a_2^2)x^2 + ...$. Substituting these series in the D.E. and collecting coefficients of like powers of x yields $(a_1^2 + a_0^2 - 1) + (4a_1 a_2 + 2a_0 a_1)x +$
$(6a_1 a_3 + 4a_2^2 + 2a_0 a_2 + a_1^2)x^2 + ... = 0$. As in the earlier problems, each coefficient must be zero. The I.C. $y(0) = 0$ requires that $a_0 = 0$, and thus $a_1^2 + a_0^2 - 1 = 0$ gives $a_1^2 = 1$. However $y'(0) = a_1 > 0$ implies $a_1 = 1$. Then $4a_1 a_2 + 2a_0 a_1 = 0$ implies that $a_2 = 0$; and

$6a_1 a_3 + 4a_2^2 + 2a_0 a_2 + a_1^2 = 6a_1 a_3 + a_1^2 = 0$ implies that $a_3 = -1/6$. Thus $y = x - x^3/3! + ...$.

23.

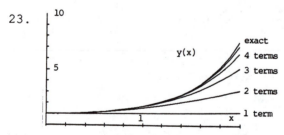

Section 5.3, Page 232

1. The D.E. can be solved for y'' to yield $y'' = -xy' - y$. If $y = \phi(x)$ is a solution, then $\phi''(x)$ $-x\phi'(x) - \phi(x)$ and thus setting $x = 0$ we obtain $\phi''(0) = - 0 - 1 = -1$. Differentiating the equation for y'' yields $y''' = -xy'' - 2y'$ and hence setting $y = \phi(x)$ again yields $\phi'''(0) = -0 - 0 = 0$. In a similar fashion $y^{iv} = -xy''' - 3y''$ and thus $\phi^{iv}(0) = - 0 - 3(-1) = 3$. The process can be continued to calculate higher derivatives of $\phi(x)$.

6. The zeros of $P(x) = x^2 - 2x - 3$ are $x = -1$ and $x = 3$. For
    $x_0 = 4$, $x_0 = -4$, and $x_0 = 0$ the distance to the nearest
    zero of $P(x)$ is $1, 3$, and 1, respectively. Thus a lower
    bound for the radius of convergence for series solutions
    in powers of $(x-4)$, $(x+4)$, and x is $\rho = 1$, $\rho = 3$, and
    $\rho = 1$, respectively.

9a. Since $P(x) = 1$ has no zeros, the radius of convergence is
    $\rho = \infty$.

9f. Since $P(x) = x^2 + 2$ has zeros at $x = \pm\sqrt{2}$, the lower
    bound for the radius of convergence of the series
    solution about $x_0 = 0$ is $\rho = \sqrt{2}$.

10a. If we assume that $y = \sum\limits_{n=0}^{\infty} a_n x^n$, then $y' = \sum\limits_{n=1}^{\infty} n a_n x^{n-1}$ and

$y'' = \sum\limits_{n=2}^{\infty} n(n-1) a_n x^{n-2}$. Substituting in the D.E., shifting
indices of summation, and collecting coefficients of like
powers of x yields the equation
$$(2\cdot 1\cdot a_2 + \alpha^2 a_0) x^0 + [3\cdot 2\cdot a_3 + (\alpha^2-1) a_1] x^1$$

$$+ \sum\limits_{n=2}^{\infty} [(n+2)(n+1) a_{n+2} + (\alpha^2-n^2) a_n] x^n = 0.$$

Hence the recurrence relation is
$a_{n+2} = (n^2-\alpha^2) a_n / (n+2)(n+1)$, $n = 0, 1, 2, \ldots$. For the
first solution we choose $a_1 = 0$. We find that
$a_2 = -\alpha^2 a_0 / 2\cdot 1$, $a_3 = 0$, $a_4 = (2^2-\alpha^2) a_2 / 4\cdot 3 = -(2^2-\alpha^2)\alpha^2 a_0 / 4!$
$\ldots$, $a_{2m} = -[(2m-2)^2 - \alpha^2] \ldots (2^2-\alpha^2)\alpha^2 a_0 / (2m)!$,

and $a_{2m+1} = 0$, so $y_1(x) = 1 - \dfrac{\alpha^2}{2!} x^2 - \dfrac{(2^2-\alpha^2)\alpha^2}{4!} x^4 - \ldots$

$$- \dfrac{[(2m-2)^2-\alpha^2] \ldots (2^2-\alpha^2)\alpha^2}{(2m)!} x^{2m} - \ldots$$

where we have set $a_0 = 1$. For the second solution we take
$a_0 = 0$ and $a_1 = 1$ in the recurrence relation to obtain
the desired solution.

10b. If α is an even integer 2k then
$(2m-2)^2 - \alpha^2 = (2m-2)^2 - 4k^2 = 0$. Thus when m = k+1 all
terms in the series for $y_1(x)$ are zero after the x^{2k} term.
A similar argument shows that if $\alpha = 2k+1$ then all terms
in $y_2(x)$ are zero after the x^{2k+1}.

11. The Taylor series about x = 0 for sinx is

$\sin x = x - x^3/3! + x^5/5! - \ldots$. Assuming that $y = \sum_{n=0}^{\infty} a_n x^n$

we find $y'' + (\sin x)y = 2a_2 + 6a_3x + 12a_4x^2 + 20a_5x^3 + 30a_6x^4$

$+ \ldots + (x-x^3/3! + \ldots)(a_0+a_1x+a_2x^2+a_3x^3+\ldots) = 2a_2 +$

$(6a_3+a_0)x + (12a_4+a_1)x^2 + (20a_5+a_2-a_0/6)x^3 +$

$(30a_6+ a_3-a_1/6)x^4 + \ldots = 0$. Hence $a_2 = 0$, $a_3 = -a_0/6$,
$a_4 = -a_1/12$, $a_5 = a_0/120$, $a_6 = (a_1+a_0)/180, \ldots$. We set
$a_0 = 1$ and $a_1 = 0$ and obtain
$y_1(x) = (1 - x^3/6 + x^5/120 + x^6/180 + \ldots)$. Next we set
$a_0 = 0$ and $a_1 = 1$ and obtain
$y_2(x) = (x - x^4/12 + x^6/180 + \ldots)$.

18. Substituting $y = \sum_{n=0}^{\infty} a_n x^n$ into the D.E. we obtain

$\sum_{n=1}^{\infty} na_n x^{n-1} - \sum_{n=0}^{\infty} a_n x^n = x^2$. Shifting indices in the

summation yields $\sum_{n=0}^{\infty} [(n+1)a_{n+1} - a_n]x^n = x^2$. Equating

coefficients of both sides then gives: $a_1 - a_0 = 0$,
$2a_2 - a_1 = 0$, $3a_3 - a_2 = 1$ and $(n+1)a_{n+1} = a_n$ for
n = 3,4,... . Thus $a_1 = a_0$, $a_2 = a_1/2 = a_0/2$, $a_3 = 1/3 +$
$a_2/3 = 1/3 + a_0/2\cdot3$, $a_4 = a_3/4 = 1/3\cdot4 + a_0/2\cdot3\cdot4$, ..., and
$a_n = a_{n-1}/n = 2/n! + a_0/n!$ and hence

$y(x) = a_0(1 + x + \dfrac{x^2}{2!}+\ldots+\dfrac{x^n}{n!}\ldots) + 2(\dfrac{x^3}{3!}+\dfrac{x^4}{4!}+\ldots+\dfrac{x^n}{n!}+\ldots)$.

Using the power series for e^x, the first and second sums
can be rewritten as desired.

20. Substituting $y = \sum_{n=0}^{\infty} a_n x^n$ into the Legendre equation,

shifting indices, and collecting coefficients of like powers of x yields

$[2 \cdot 1 \cdot a_2 + \alpha(\alpha+1) a_0] x^0 + \{3 \cdot 2 \cdot a_3 - [2 \cdot 1 - \alpha(\alpha+1)] a_1\} x^1 +$

$\sum_{n=2}^{\infty} \{(n+2)(n+1) a_{n+2} - [n(n+1) - \alpha(\alpha+1)] a_n\} x^n = 0.$ Thus

$a_2 = -\alpha(\alpha+1) a_0/2!$, $a_3 = [2 \cdot 1 - \alpha(\alpha+1)] a_1/3! =$
$-(\alpha-1)(\alpha+2) a_1/3!$ and the recurrence relation is
$(n+2)(n+1) a_{n+2} = -[\alpha(\alpha+1) - n(n+1)] a_n = -(\alpha-n)(\alpha+n+1) a_n$,
$n = 2, 3, \ldots$. Setting $a_1 = 0$, $a_0 = 1$ yields a solution
with $a_3 = a_5 = a_7 = \ldots = 0$ and
$a_4 = \alpha(\alpha-2)(\alpha+1)(\alpha+3)/4!, \ldots, a_{2m} = (-1)^m \alpha(\alpha-2)(\alpha-4) \ldots$
$(\alpha-2m+2)(\alpha+1)(\alpha+3) \ldots (\alpha+2m-1)/(2m)!, \ldots$. The second
(linearly independent) solution is obtained by setting
$a_0 = 0$ and $a_1 = 1$. The coefficients are $a_2 = a_4 = a_6 =$
$\ldots = 0$ and $a_3 = -(\alpha-1)(\alpha+2)/3!$, $a_5 = -(\alpha-3)(\alpha+4) a_3/5 \cdot 4 =$
$(\alpha-1)(\alpha-3)(\alpha+2)(\alpha+4)/5!, \ldots$.

25. Using the chain rule we have:
$$\frac{dF(\phi)}{d\phi} = \frac{dF[\phi(x)]}{dx} \frac{dx}{d\phi} = -f'(x) \sin\phi(x) = -f'(x)\sqrt{1-x^2},$$
$$\frac{d^2F(\phi)}{d\phi^2} = \frac{d}{dx}[-f'(x)\sqrt{1-x^2}] \frac{dx}{d\phi} = (1-x^2) f''(x) - xf'(x),$$
which when substituted into the D.E. yields the desired result.

27. Carrying out the steps indicated yields the two equations:
$$P_m[(1-x^2) P_n']' = -n(n+1) P_n P_m$$
$$P_n[(1-x^2) P_m']' = -m(m+1) P_n P_m.$$
As long as $n \neq m$ the second equation can be subtracted from the first and the result integrated from -1 to 1 to obtain
$$\int_{-1}^{1} \{P_m[(1-x^2) P_n']' - P_n[(1-x^2) P_m']'\} dx = [m(m+1) - n(n+1)] \int_{-1}^{1} P_n P_m dx.$$

The left side may be integrated by parts to yield

$$[P_m(1-x^2)P_n' - P_n(1-x^2)P_m']_{-1}^{1} + \int_{-1}^{1}[P_m'\,(1-x^2)P_n' - P_n'(1-x^2)P_m']dx,$$

which is zero. Thus $\int_{-1}^{1}P_n(x)P_m(x)dx = 0$ for $n \neq m$.

Section 5.4, Page 238

1. Since the coefficients of y, y' and y'' have no common
    factors and since $P(x)$ vanishes only at $x = 0$ we conclude
    that $x = 0$ is a singular point. Writing the D.E. in the
    form $y'' + p(x)y' + q(x)y = 0$, we obtain $p(x) = (1-x)/x$
    and $q(x) = 1$. Thus for the singular point we have
    $\lim_{x\to 0} xp(x) = \lim_{x\to 0} 1-x = 1$, $\lim_{x\to 0} x^2q(x) = 0$ and thus $x = 0$ is
    a regular singular point.

5. Writing the D.E. in the form $y'' + p(x)y' + q(x)y = 0$, we
    find $p(x) = x/(1-x)(1+x)^2$ and $q(x) = 1/(1-x^2)(1+x)$.
    Therefore $x = \pm 1$ are singular points. Since
    $\lim_{x\to 1} (x-1)p(x)$ and $\lim_{x\to 1} (x-1)^2q(x)$ both exist, we conclude
    $x = 1$ is a regular singular point. Finally, since
    $\lim_{x\to -1} (x+1)p(x)$ does not exist, we find that $x = -1$ is an
    irregular singular point.

12. Writing the D.E. in the form $y + p(x)y' + q(x)y = 0$, we
    see that $p(x) = e^x/x$ and $q(x) = (3\cos x)/x$. Thus $x = 0$ is
    a singular point. Since $xp(x) = e^x$ is analytic at $x = 0$
    and $x^2q(x) = 3x\cos x$ is analytic at $x = 0$ the point $x = 0$
    is a regular singular point.

17. Writing the D.E. in the form $y'' + p(x)y' + q(x)y = 0$, we
    see that $p(x) = \dfrac{x}{\sin x}$ and $q(x) = \dfrac{4}{\sin x}$. Since $\lim_{x\to 0} q(x)$
    does not exist the point $x_0 = 0$ is a singular point and
    since neither $\lim_{x\to \pm n\pi} p(x)$ nor $\lim_{x\to \pm n\pi} q(x)$ exist either the
    points $x_0 = \pm n\pi$ are also singular points. To determine
    whether the singular points are regular or irregular we
    must use Eq.(8) and the result #7 of multiplication and
    division of power series from Section 5.1. For $x_0 = 0$,
    we have

$$xp(x) = \frac{x^2}{\sin x} = \frac{x^2}{x - \frac{x^3}{6} + \cdots} = x[1 + \frac{x^2}{6} + \cdots]$$

$$= x + \frac{x^3}{6} + \cdots,$$

which converges about $x_0 = 0$ and thus $xp(x)$ is analytic at $x_0 = 0$. $x^2q(x)$, by similar steps, is also analytic at $x_0 = 0$ and thus $x_0 = 0$ is a regular singular point. For $x_0 = n\pi$, we have

$$(x-n\pi)p(x) = \frac{(x-n\pi)x}{\sin x} = \frac{(x-n\pi)[(x-n\pi) + n\pi]}{\pm(x-n\pi) + \frac{-(x-n\pi)^3}{6} \pm \cdots}$$

$$= [(x-n\pi)+n\pi][\pm 1 + \frac{-(x-n\pi)^2}{6} \pm \cdots], \text{ which}$$

converges about $x_0 = n\pi$ and thus $(x-n\pi)p(x)$ is analytic at $x = n\pi$. Similarly $(x+n\pi)p(x)$ and $(x\pm n\pi)^2 q(x)$ are analytic and thus $x_0 = \pm n\pi$ are regular singular points.

19. Substituting $y = \sum_{n=0}^{\infty} a_n x^n$ into the D.E. yields

$$2\sum_{n=2}^{\infty} n(n-1)a_n x^{n-1} + 3\sum_{n=1}^{\infty} na_n x^{n-1} + \sum_{n=0}^{\infty} a_n x^{n+1} = 0. \quad \text{The last sum}$$

becomes $\sum_{n=2}^{\infty} a_{n-2} x^{n-1}$ by replacing $n+1$ by $n-1$, the first term of the middle sum is $3a_1$, and thus we have

$$3a_1 + \sum_{n=2}^{\infty} \{[2n(n-1)+3n]a_n + a_{n-2}\}x^{n-1} = 0. \quad \text{Hence } a_1 = 0 \text{ and}$$

$$a_n = \frac{-a_{n-2}}{n(2n+1)}, \text{ which is the desired recurrance relation.}$$

Thus all even coefficients are found in terms of a_0 and all odd coefficients are zero, thereby yielding only one solution of the desired form.

21. If $\xi = 1/x$ then

$$\frac{dy}{dx} = \frac{dy}{d\xi}\frac{d\xi}{dx} = -\frac{1}{x^2}\frac{dy}{d\xi} = -\xi^2\frac{dy}{d\xi},$$

$$\frac{d^2y}{dx^2} = \frac{d}{d\xi}(-\xi^2\frac{dy}{d\xi})\ \frac{d\xi}{dx} = (-2\xi\ \frac{dy}{d\xi} - \xi^2\frac{d^2y}{d\xi^2})\ (-\frac{1}{x^2})$$

$$= \xi^4\frac{d^2y}{d\xi^2} + 2\xi^3\frac{dy}{d\xi}.$$

Substituting in the D.E. we have

$$P(1/\xi)\ [\xi^4\frac{d^2y}{d\xi^2} + 2\xi^3\frac{dy}{d\xi}] + Q(1/\xi)\ [-\xi^2\frac{dy}{d\xi}] + R(1/\xi)y = 0,$$

$$\xi^4P(1/\xi)\frac{d^2y}{d\xi^2} + [2\xi^3P(1/\xi) - \xi^2Q(1/\xi)]\frac{dy}{d\xi} + R(1/\xi)y = 0.$$

The result then follows from the theory of singular points at $\xi = 0$.

23. Since $P(x) = x^2$, $Q(x) = x$, and $R(x) = x^2 - v^2$, $f(\xi) =$
    $[2P(1/\xi)/\xi - Q(1/\xi)/\xi^2]/P(1/\xi) = 2/\xi - 1/\xi = 1/\xi$ and
    $g(\xi) = R(1/\xi)/\xi^4P(1/\xi) = (1/\xi^2 - v^2)/\xi^2 = 1/\xi^4 - v^2/\xi^2$.
    Thus the point at infinity is a singular point. Although
    $\xi f(\xi) = 1$ is analytic at $\xi = 0$, $\xi^2g(\xi) = 1/\xi^2 - v^2$ is not,
    so the point at infinity is an irregular singular point.

Section 5.5, Page 246

1. Assume $y = (x+1)^r$ for $x + 1 > 0$. Substitution of y into
    the D.E. yields $[r(r-1) + 3r + 3/4](x+1)^r = 0$. Thus
    $r^2 + 2r + 3/4 = 0$, which yields $r = -3/2, -1/2$. The
    general solution of the D.E. is then
    $y = c_1|x+1|^{-1/2} + c_2|x+1|^{-3/2}$, $x \neq -1$.

4. If $y = x^r$ then $r(r-1) + 3r + 5 = 0$. So $r^2 + 2r + 5 = 0$
    and $r = (-2 \pm \sqrt{4-20})/2 = -1 \pm 2i$. Thus the general
    solution of the D.E. is
    $y = c_1x^{-1}\cos(2\ln|x|) + c_2x^{-1}\sin(2\ln|x|)$, $x \neq 0$.

9. Again let $y = x^r$ to obtain $r(r-1) - 5r + 9 = 0$, or
    $(r-3)^2 = 0$. Thus the roots are $x = 3,3$ and
    $y = c_1x^3 + c_2x^3\ln|x|$, $x \neq 0$, is the solution of the D.E.

13. If $y = x^r$, then $F(r) = 2r(r-1) + r - 3 = 2r^2 - r - 3 =$
    $(2r-3)(r+1) = 0$, so $y = c_1x^{3/2} + c_2x^{-1}$ and

$y' = \dfrac{3}{2}c_1x^{1/2} - c_2x^{-2}$. Setting $x = 1$ in y and y' we obtain

$c_1 + c_2 = 1$ and $\dfrac{3}{2}c_1 - c_2 = 4$, which yield $c_1 = 2$ and

$c_2 = -1$. Hence $y = 2x^{3/2} - x^{-1}$.

17. Substituting $y = x^r$, we find that $r(r-1) + \alpha r + 5/2 = 0$
or $r^2 + (\alpha-1)r + 5/2 = 0$. Thus
$r_1, r_2 = [-(\alpha-1) \pm \sqrt{(\alpha-1)^2-10}]\,/2$. In order for solutions
to approach zero as $x \to 0$ it is necessary that the real
parts of r_1 and r_2 be positive. Suppose that $\alpha > 1$, then
$\sqrt{(\alpha-1)^2-10}$ is either imaginary or real and less than
$\alpha - 1$; hence the real parts of r_1 and r_2 will be
negative. Suppose that $\alpha = 1$, then $r_1, r_2 = \pm i\sqrt{10}$ and
the solutions are oscillatory. Suppose that $\alpha < 1$, then
$\sqrt{(\alpha-1)^2-10}$ is either imaginary or real and less than
$|\alpha-1| = 1 - \alpha$; hence the real parts of r_1 and r_2 will be
positive. Thus if $\alpha < 1$ the solutions of the D.E. will
approach zero as $x \to 0$.

21. In all cases the roots of $F(r) = 0$ are given by Eq.(5)
and the forms of the solution are given in Theorem 5.5.1.

22. Assume that $y = v(x)x^{r_1}$. Then $y' = v(x)r_1x^{r_1-1} + v'(x)x^{r_1}$
and $y'' = v(x)r_1(r_1-1)x^{r_1-2} + 2v'(x)r_1x^{r_1-1} + v''(x)x^{r_1}$.
Substituting in the D.E. and collecting terms yields
$x^{r_1+2}v'' + (\alpha + 2r_1)x^{r_1+1}v' + [r_1(r_1-1) + \alpha r_1 + \beta]x^{r_1}v = 0$.
Now we make use of the fact that r_1 is a double root of
$f(r) = r(r-1) + \alpha r + \beta$. This means that $f(r_1) = 0$ and
$f'(r_1) = 2r_1 - 1 + \alpha = 0$. Hence the D.E. for v reduces
to $x^{r_1+2}v'' + x^{r_1+1}v'$. Since $x > 0$ we may divide by x^{r_1+1}
to obtain $xv'' + v' = 0$. Thus $v(x) = \ln x$ and a second
solution is $y = x^{r_1}\ln x$.

25. The change of variable $x = e^z$ transforms the D.E. into
$u'' - 4u' + 4u = z$, which has the solution
$u(z) = c_1e^{2z} + c_2ze^{2z} + (1/4)z + 1/4$. Hence
$y(x) = c_1x^2 + c_2x^2\ln x + (1/4)\ln x + 1/4$.

31. If $x > 0$, then $|x| = x$ and $|x|^{r_1} = x^{r_1}$ so we can choose

$c_1 = k_1$. If $x < 0$, then $|x| = -x$ and $|x|^{r_1} = (-x)^{r_1} = (-1)^{r_1}x^{r_1}$ and we can choose $c_1 = (-1)^{r_1}k_1$, or $k_1 = (-1)^{r_1}c_1$. In both bases we have $c_2 = k_2$.

33.   If $y = x^r$ then $r(r-1) + (1-i)r + 2 = 0$. Simplifying, we have $r^2 - ir + 2 = (r-2i)(r+i) = 0$, so $r = -i$ and $r = 2i$. If one does not notice that the quadratic can be factored, then the quadratic formula can be used: $r = (i \pm \sqrt{-1-8})/2 = (i \pm 3i)/2 = -i$ and $2i$. The general solution of the D.E. is $y = c_1x^{2i} + c_2x^{-i}$, $x \neq 0$. Note that the solutions are complex-valued functions of the real variable x.

Section 5.6, Page 251

2.   If the D.E. is put in the standard form $y'' + p(x)y + q(x)y = 0$, then $p(x) = x^{-1}$ and $q(x) = 1 - 1/9x^2$. Thus $x = 0$ is a singular point. Since $xp(x) \to 1$ and $x^2q(x) \to -1/9$ as $x \to 0$ it follows that $x = 0$ is a regular singular point. In determining a series solution of the D.E. it is more convenient to leave the equation in the form given rather than divide by the x^2, the coefficient of y''. If we substitute $y = \sum_{n=0}^{\infty}a_nx^{n+r}$, we have

$$\sum_{n=0}^{\infty} (n+r)(n+r-1)a_nx^{n+r} + \sum_{n=0}^{\infty} (n+r)a_nx^{n+r} + (x^2 - \frac{1}{9})\sum_{n=0}^{\infty} a_nx^{n+r} = 0.$$

Note that $x^2 \sum_{n=0}^{\infty} a_nx^{n+r} = \sum_{n=0}^{\infty} a_nx^{n+r+2} = \sum_{n=2}^{\infty} a_{n-2}x^{n+r}$. Thus we have $[r(r-1) + r - \frac{1}{9}]a_0x^r + [(r+1)r + (r+1) - \frac{1}{9}]a_1x^{r+1} +$

$$\sum_{n=2}^{\infty} \{[(n+r)(n+r-1) + (n+r) - \frac{1}{9}]a_n + a_{n-2}\} x^{n+r} = 0.$$ The indicial equation is $r^2 - 1/9 = 0$ with roots $r_1 = 1/3$ and $r_2 = -1/3$. For either value of r it is necessary to take $a_1 = 0$ in order that the coefficient of x^{r+1} be zero. The recurrence relation is $[(n+r)^2 - 1/9]a_n = -a_{n-2}$. For $r = 1/3$ we have

$$a_n = \frac{-a_{n-2}}{(n + \frac{1}{3})^2 - (\frac{1}{3})^2} = -\frac{a_{n-2}}{(n + \frac{2}{3})n}, \quad n = 2,3,4,\ldots .$$

Since $a_1 = 0$ it follows from the recurrence relation that $a_3 = a_5 = a_7 = \ldots = 0$. For the even coefficients it is convenient to let $n = 2m$, $m = 1,2,3,\ldots$. Then $a_{2m} = -a_{m-2}/2^2 m(m + \frac{1}{3})$. The first few coefficients are given by

$$a_2 = \frac{(-1)a_0}{2^2(1 + \frac{1}{3})1}, \quad a_4 = \frac{(-1)a_2}{2^2(2 + \frac{1}{3})2} = \frac{a_0}{2^4(1 + \frac{1}{3})(2 + \frac{1}{3})2!}$$

$$a_6 = \frac{(-1)a_4}{2^2(3 + \frac{1}{3})3} = \frac{(-1)a_0}{2^6(1 + \frac{1}{3})(2 + \frac{1}{3})(3 + \frac{1}{3})3!}, \text{ and the}$$

coefficent of x^{2m} for $m = 1, 2,\ldots$ is

$$a_{2m} = \frac{(-1)^m a_0}{2^{2m} m!(1 + \frac{1}{3})(2 + \frac{1}{3}) \ldots (m + \frac{1}{3})}. \quad \text{Thus one}$$

solution (on setting $a_0 = 1$) is

$$y_1(x) = x^{1/3}[1 + \sum_{m=1}^{\infty} \frac{(-1)^m}{m! (1 + \frac{1}{3})(2 + \frac{1}{3})\ldots(m + \frac{1}{3})} (\frac{x}{2})^{2m}].$$

Since $r_2 = -1/3 \neq r_1$ and $r_1 - r_2 = 2/3$ is not an integer, we can calculate a second series solution corresponding to $r = -1/3$. The recurrence relation is $n(n-2/3)a_n = -a_{n-2}$, which yields the desired solution following the steps just outlined. Note that $a_1 = 0$, as in the first solution, and thus all the odd coefficients are zero.

4. Putting the D.E. in standard form $y'' + p(x)y' + q(x)y = 0$, we see that $p(x) = 1/x$ and $q(x) = -1/x$. Thus $x = 0$ is a singular point, and since $xp(x) \to 1$ and $x^2 q(x) \to 0$, as $x \to 0$, $x = 0$ is a regular singular point. Substituting

$$y = \sum_{n=0}^{\infty} a_n x^{n+r} \text{ in } xy'' + y' - y = 0 \text{ and shifting indices we}$$

obtain

$$\sum_{n=-1}^{\infty} a_{n+1}(r+n+1)(r+n)x^{n+r} + \sum_{n=-1}^{\infty} a_{n+1}(r+n+1)x^{n+r} - \sum_{n=-}^{\infty} a_n x^{n+r} = 0,$$

$$[r(r-1) + r]a_0 x^{-1+r} + \sum_{n=0}^{\infty} [(r+n+1)^2 a_{n+1} - a_n]x^{n+r} = 0. \quad \text{The}$$

indicial equation is $r^2 = 0$ so $r = 0$ is a double root.
Thus we will obtain only one series of the form

$$y = x^r \sum_{n=0}^{\infty} a_n x^n. \quad \text{The recurrence relation is}$$

$(n+1)^2 a_{n+1} = a_n$, $n = 0,1,2,\ldots$. The coefficients are
$a_1 = a_0$, $a_2 = a_1/2^2 = a_0/2^2$, $a_3 = a_2/3^2 = a_0/3^2 \cdot 2^2$,
$a_4 = a_3/4^2 = a_0/4^2 \cdot 3^2 \cdot 2^2, \ldots$ and $a_n = a_0/(n!)^2$. Thus one

solution (on setting $a_0 = 1$) is $y = \displaystyle\sum_{n=0}^{\infty} x^n/(n!)^2$.

11. If we make the change of variable $z = x-1$ and let
$y = u(z)$, then the Legendre equation transforms to
$(z^2 + 2z)u''(z) + 2(z+1)u'(z) - \alpha(\alpha+1)u(z) = 0$. Since
$x = 1$ is a regular singular point of the original
equation, we know that $z = 0$ is a regular singular point

of the transformed equation. Substituting $u = \displaystyle\sum_{n=0}^{\infty} a_n z^{n+r}$

in the transformed equation and shifting indices, we
obtain

$$\sum_{n=0}^{\infty} (n+r)(n+r-1)a_n z^{n+r} + 2\sum_{n=-1}^{\infty} (n+r+1)(n+r)a_{n+1}z^{n+r}$$

$$+ 2\sum_{n=0}^{\infty} (n+r)a_n z^{n+r} + 2\sum_{n=-1}^{\infty} (n+r+1)a_{n+1}z^{n+r} - \alpha(\alpha+1)\sum_{n=0}^{\infty}$$

$$a_n z^{n+r} = 0, \quad \text{or}$$

$$[2r(r-1) + 2r]a_0 + \sum_{n=0}^{\infty} \{2(n+r+1)^2 a_{n+1}$$

$$+ [(n+r)(n+r+1) - \alpha(\alpha+1)]a_n\}z^{n+r} = 0.$$

The indicial equation is $2r^2 = 0$ so $r = 0$ is a double
root. Thus there will be only one series solution of the

form $y = \sum_{n=0}^{\infty} a_n z^{n+r}$. The recurrence relation is

$2(n+1)^2 a_{n+1} = [\alpha(\alpha+1) - n(n+1)]a_n, n = 0,1,2,\ldots$. We have

$a_1 = [\alpha(\alpha+1)]a_0/2\cdot 1^2$, $a_2 = [\alpha(\alpha+1)][\alpha(\alpha+1) - 1\cdot 2]a_0/2^2\cdot 2^2\cdot 1^2$,

$a_3 = [\alpha(\alpha+1)][\alpha(\alpha+1) - 1\cdot 2][\alpha(\alpha+1) - 2\cdot 3]a_0/2^3\cdot 3^2\cdot 2^2\cdot 1^2, \ldots,$

and $a_n = [\alpha(\alpha+1)][\alpha(\alpha+1)-1\cdot 2]\ldots[\alpha(\alpha+1)-(n-1)n]a_0/2^n(n!)^2$.

Reverting to the variable x it follows that one solution
of the Legendre equation in powers of x-1 is

$y_1(x) = \sum_{n=0}^{\infty} [\alpha(\alpha+1)][\alpha(\alpha+1) - 1\cdot 2] \ldots$

$[\alpha(\alpha+1) - (n-1)n][(x-1)^n/2^n(n!)]^2$ where we have set $a_0 = 1$.

14. The standard form is $y'' + p(x)y' + q(x)y = 0$, with
$p(x) = 1/x$ and $q(x) = 1$. Thus $x = 0$ is a singular point;
and since $xp(x) \to 1$ and $x^2 q(x) \to 0$ as $x \to 0$, $x = 0$ is a

regular singular point. Substituting $y = \sum_{n=0}^{\infty} a_n x^{n+r}$ into

$x^2 y'' + xy' + x^2 y = 0$ and shifting indices appropriately,
we obtain

$\sum_{n=0}^{\infty} (n+r)(n+r-1)a_n x^{n+r} + \sum_{n=0}^{\infty} (n+r)a_n x^{n+r} + \sum_{n=2}^{\infty} a_{n-2} x^{n+r} = 0,$

$[r(r-1)+r]a_0 x^r + [(1+r)r+1+r]a_1 x^{r+1}$

$+\sum_{n=2}^{\infty} [(n+r)^2 a_n + a_{n-2}]x^{n+r} = 0.$ The indicial equation is

$r^2 = 0$ so $r = 0$ is a double root. It is necessary to
take $a_1 = 0$ in order that the coefficient of x^{r+1} be zero.
The recurrence relation in $n^2 a_n = -a_{n-2}, n = 2,3,\ldots$.
Since $a_1 = 0$ it follows that $a_3 = a_5 = a_7 = \ldots = 0$. For
the even coefficients we let $n = 2m$, $m = 1,2,\ldots$. Then
$a_{2m} = -a_{2m-2}/2^2 m^2$ so $a_2 = -a_0/2^2\cdot 1^2$, $a_4 = a_0/2^2\cdot 2^2\cdot 1^2\cdot 2^2, \ldots$,
and $a_{2m} = (-1)^m a_0/2^{2m}(m!)^2$. Thus one solution of the Bessel

equation of order zero is $J_0(x) = 1 + \sum_{m=1}^{\infty} (-1)^m x^{2m}/2^{2m}(m!)^2$

where we have set $a_0 = 1$. Using the ratio test it can be
shown that the series converges for all x. Also note
that $J_0(x) \to 1$ as $x \to 0$.

15. In order to determine the form of the integral for x near
zero we must study the integrand for x small. Using the
above series for J_0, we have

$$\frac{1}{x[J_0(x)]^2} = \frac{1}{x[1 - x^2/2 + \ldots]^2} = \frac{1}{x[1 - x^2 + \ldots]} =$$

$$\frac{1}{x}[1 + x^2 + \ldots]\text{ for x small.} \quad \text{Thus}$$

$$y_2(x) = J_0(x)\int \frac{dx}{x[J_0(x)]^2} = J_0(x)\int [\frac{1}{x} + x + \ldots]dx$$

$$= J_0(x)[\ln x + \frac{x^2}{2} + \ldots],$$

and it is clear that $y_2(x)$ will contain a logarithmic
term.

16a. Putting the D.E. in the standard form
$y'' + p(x)y' + q(x)y = 0$ we see that $p(x) = 1/x$ and
$q(x) = (x^2-1)/x^2$. Thus $x = 0$ is a singular point and
since $xp(x) \to 1$ and $x^2q(x) \to -1$ as $x \to 0$, $x = 0$ is a

regular singular point. Substituting $y = \sum_{n=0}^{\infty} a_n x^{n+r}$ into

$x^2y'' + xy' + (x^2-1)y = 0$, shifting indices appropriately,
and collecting coefficients of common powers of x we
obtain $[r(r-1) + r - 1]a_0 x^r + [(1+r)r + 1 + r - 1]a_1 x^{r+1}$

$$+ \sum_{n=2}^{\infty} \{[(n+r)^2 - 1]a_n + a_{n-2}\}x^{n+r} = 0.$$

The indicial equation is $r^2-1 = 0$ so the roots are $r_1 = 1$
and $r_2 = -1$. For either value of r it is necessary to
take $a_1 = 0$ in order that the coefficient of x^{r+1} be zero.
The recurrence relation is $[(n+r)^2 - 1]a_n = -a_{n-2}$,
$n = 2,3,4\ldots$. For $r = 1$ we have $a_n = -a_{n-2}/[n(n+2)]$,
$n = 2,3,4,\ldots$. Since $a_1 = 0$ it follows that $a_3 = a_5 = a_7$
$= \ldots = 0$. Let $n = 2m$. Then $a_{2m} = -a_{2m-2}/2^2m(m+1)$,
$m = 1,2,\ldots$, so $a_2 = -a_0/2^2 \cdot 1 \cdot 2$, $a_4 = -a_2/2^2 \cdot 1 \cdot 2 \cdot 3 =$
$a_0/2^2 \cdot 2^2 \cdot 1 \cdot 2 \cdot 2 \cdot 3,\ldots$, and $a_{2m} = (-1)^m a_0/2^{2m}m!(m+1)!$. Thus one
solution (set $a_0 = 1/2$) of the Bessel equation of order

one is $J_1(x) = (x/2) \sum_{n=0}^{\infty} (-1)^n x^{2n}/(n+1)!n!2^{2n}$. The ratio

test shows that the series converges for all x. Also

note that $J_1(x) \to 0$ as $x \to 0$.

16b. For $r = -1$ the recurrence relation is
$[(n-1)^2 - 1]a_n = -a_{n-2}$, $n = 2,3,\ldots$. Substituting $n = 2$
into the relation yields $[(2-1)^2 - 1]a_2 = 0$ $a_2 = -a_0$.
Hence it is impossible to determine a_2 and consequently
impossible to find a series solution of the form

$$x^{-1} \sum_{n=0}^{\infty} b_n x^n.$$

Secton 5.7, Page 257

1. The D.E. has the form $P(x)y'' + Q(x)y' + R(x)y = 0$ with
$P(x) = x$, $Q(x) = 2x$, and $R(x) = 6e^x$. From this we find
$p(x) = Q(x)/P(x) = 2$ and $q(x) = R(x)/P(x) = 6e^x/x$ and
thus $x = 0$ is a singular point. Since $xp(x) = 2x$ and
$x^2 q(x) = 6xe^x$ are analytic at $x = 0$ we conclude that
$x = 0$ is a regular singular point. Next, we have
$xp(x) \to 0 = p_0$ and $x^2 q(x) \to 0 = q_0$ as $x \to 0$ and thus the
indicial equation is $r(r-1) + 0 \cdot r + 0 = r^2 - r = 0$, which
has the roots $r_1 = 1$ and $r_2 = 0$.

3. The equation has the form $P(x)y'' + Q(x)y' + R(x)y = 0$
with $P(x) = x(x-1)$, $Q(x) = 6x^2$ and $R(x) = 3$. Since $P(x)$,
$Q(x)$, and $R(x)$ are polynomials with no common factors and
$P(0) = 0$ and $P(1) = 0$, we conclude that $x = 0$ and $x = 1$
are singular points. The first point, $x = 0$, can be shown
to be a regular singular point using steps similar to
those to shown in Problem 1. For $x = 1$, we must put the
D.E. in a form similar to Eq.(1) for this case. To do
this, divide the D.E. by x and multiply by $(x-1)$ to
obtain $(x-1)^2 y'' + 6x(x-1)y + \dfrac{3}{x}(x-1)y = 0$. Comparing this
to Eq.(1) we find that $(x-1)p(x) = 6x$ and
$(x-1)^2 q(x) = 3(x-1)/x$ which are both analytic at
$x = 1$ and hence $x = 1$ is a regular singular point. These
last two expressions approach $p_0 = 6$ and $q_0 = 0$
respectively as $x \to 1$, and thus the indicial equation is
$r(r-1) + 6r + 0 = r(r+5) = 0$.

9. For this D.E., $p(x) = \dfrac{-(1+x)}{x^2(1-x)}$ and $q(x) = \dfrac{2}{x(1-x)}$ and thus

$x = 0$, -1 are singular points. Since $xp(x)$ is not analytic at $x = 0$, $x = 0$ is not a regular singular point. Looking at $(x-1)p(x) = \dfrac{1+x}{x^2}$ and $(x-1)^2 q(x) = \dfrac{2(1-x)}{x}$ we see that $x = 1$ is a regular singular point and that $p_0 = 2$ and $q_0 = 0$.

13. We have $p(x) = \dfrac{\sin x}{x^2}$ and $q(x) = -\dfrac{\cos x}{x^2}$, so that $x = 0$ is a singular point. Note that $xp(x) = (\sin x)/x \to 1 = p_0$ as $x \to 0$ and $x^2 q(x) = -\cos x \to -1 = q_0$ as $x \to 0$. In order to assert that $x = 0$ is a regular singular point we must demonstrate that $xp(x)$ and $x^2 q(x)$, with $xp(x) = 1$ at $x = 0$ and $q(x) = -1$ at $x = 0$, have convergent power series (are analytic) about $x = 0$. We know that $\cos x$ is analytic so we need only consider $(\sin x)/x$. But

$$\sin x = \sum_{n=0}^{\infty} x^{2n+1}/(2n+1)! \text{ for } -\infty < x < \infty \text{ so}$$

$$(\sin x)/x = \sum_{n=0}^{\infty} x^{2n}/(2n+1)! \text{ and hence is analytic.} \text{ We}$$

conclude that $x = 0$ is a regular singular point. It follows that the indicial equation is $r(r-1) + r - 1 = r^2 - 1 = 0$ and the roots are $r_1 = 1$, $r_2 = -1$. To find the first few terms of the solution corresponding to $r_1 = 1$, assume that
$y = x(a_0 + a_1 x + a_2 x^2 + \dots) = a_0 x + a_1 x^2 + a_2 x^3 + \dots$.
Substituting this series for y in the D.E. and expanding $\sin x$ and $\cos x$ about $x = 0$ yields
$x^2(2a_1 + 6a_2 x + 12a_3 x^2 + 20a_4 x^3 + \dots) +$
$(x - x^3/3! + x^5/5! + \dots)(a_0 + 2a_1 x + 3a_2 x^2 + 4a_3 x^3 + 5a_4 x^4 + \dots) - (1 - x^2/2! + x^4/4! - \dots)(a_0 x + a_1 x^2 + a_2 x^3 + a_3 x^4 + a_4 x^5 + \dots) = 0$. Collecting terms, $(2a_1 + 2a_1 - a_1)x^2 + (6a_2 + 3a_2 - a_0/6 - a_2 + a_0/2)x^3 + (12a_3 + 4a_3 - 2a_1/6 - a_3 + a_1/2)x^4 + (20a_4 + 5a_4 - 3a_2/6 + a_0/120 - a_4 + a_2/2 - a_0/24)x^5 + \dots = 0$. Simplifying, $3a_1 x^2 + (8a_2 + a_0/3)x^3 + (15a_3 + a_1/6)x^4 + (24a_4 - a_0/30)x^5 + \dots = 0$. Thus, $a_1 = 0$, $a_2 = -a_0/4!$, $a_3 = 0$, $a_4 = a_0/6!,\dots$. Hence
$y_1(x) = x - x^3/4! + x^5/6! + \dots$ where we have set $a_0 = 1$.

14. We first write the D.E. in the standard form as given for Theorem 5.7.1 except that we are expanding in powers of (x-1) rather than powers of x:

$(x-1)^2 y'' + (x-1)[(x-1)/2\ln x]y' + [(x-1)^2/\ln x]y = 0$. since $\ln 1 = 0$, $x = 1$ is a singular point. To show it is a regular singular point of this D.E. we must show that $(x-1)/\ln x$ is analytic at $x = 1$; it will then follow that $(x-1)^2/\ln x = (x-1)[(x-1)/\ln x]$ is also analytic at $x = 1$. If we expand $\ln x$ in a Taylor series about $x = 1$ we find that $\ln x = (x-1) - \dfrac{1}{2}(x-1)^2 + \dfrac{1}{3}(x-1)^3 - \cdots$.

Thus $(x-1)/\ln x = [1 - \dfrac{1}{2}(x-1) + \dfrac{1}{3}(x-1)^2 - \cdots]^{-1} =$

$1 + \dfrac{1}{2}(x-1) + \cdots$ has a power series expansion about $x = 1$, and hence is analytic. We can use the above result to obtain the indicial equation at $x = 1$. We have

$(x-1)^2 y'' + (x-1)[\dfrac{1}{2} + \dfrac{1}{4}(x-1) + \cdots]y' + [(x-1) +$

$\dfrac{1}{2}(x-1)^2 + \cdots]y = 0$. Thus $p_0 = 1/2$, $q_0 = 0$ and the indicial equation is $r(r-1) + r/2 = 0$. Hence $r = 1/2$ and $r = 0$. In order to find the first three non-zero terms in a series solution corresponding to $r = 1/2$, it is better to keep the differential equation in its original form and to substitute the above power series for $\ln x$:

$[(x-1) - \dfrac{1}{2}(x-1)^2 + \dfrac{1}{3}(x-1)^3 - \dfrac{1}{4}(x-1)^4 + \cdots]y'' + \dfrac{1}{2}y' + y = 0$.

Next we substitute $y = a_0(x-1)^{1/2} + a_1(x-1)^{3/2} + a_2(x-1)^{5/2} + \cdots$ and collect coefficients of like powers of (x-1) which are then set equal to zero. This requires some algebra before we find that $6a_1/4 + 9a_0/8 = 0$ and $5a_2 + 5a_1/8 - a_0/12 = 0$. These equations yield $a_1 = -3a_0/4$ and $a_2 = 53a_0/480$. With $a_0 = 1$ we obtain the solution

$y_1(x) = (x-1)^{1/2} - \dfrac{3}{4}(x-1)^{3/2} + \dfrac{53}{480}(x-1)^{5/2} + \cdots$. Since the radius of convergence of the series for $\ln x$ is 1, we would expect $\rho = 1$.

16a. If we write the D.E. in the standard form as given in Theorem 5.7.1 we obtain $x^2 y'' + x[\alpha/x]y' + [\beta/x]y = 0$ where $xp(x) = \alpha/x$ and $x^2 q(x) = \beta/x$. Neither of these terms are analytic at $x = 0$ so $x = 0$ is an irregular singular point.

16b. Substituting $y = x^r \sum\limits_{n=0}^{\infty} a_n x^n$ in $x^3 y'' + \alpha x y' + \beta y = 0$ gives

$$\sum_{n=0}^{\infty} (n+r)(n+r-1) a_n x^{n+r+1} + \alpha \sum_{n=0}^{\infty} (n+r) a_n x^{n+r} + \beta \sum_{n=0}^{\infty} a_n x^{n+r}.$$

Shifting the index in the first series and collecting coefficients of common powers of x we obtain $(\alpha r + \beta) a_0 x^r$

$$+ \sum_{n=1}^{\infty} (n+r-1)(n+r-2) a_{n-1} + [\alpha(n+r) + \beta] a_n x^{n+r} = 0. \quad \text{Thus}$$

the indicial equation is $\alpha r + \beta = 0$ with the single root $r = -\beta/\alpha$.

16c. From part b, the recurrence relation is

$$a_n = \frac{(n+r-1)(n+r-2) a_{n-1}}{\alpha(n+r) + \beta}, \quad n = 1, 2, \ldots$$

$$= \frac{(n - \frac{\beta}{\alpha} - 1)(n - \frac{\beta}{\alpha} - 2) a_{n-1}}{\alpha n}.$$

For $\frac{\beta}{\alpha} = -1$, then, $a_n = \frac{n(n-1) a_{n-1}}{\alpha n}$, which is zero for $n = 1$ and thus $y(x) = x$ is the solution. Similarly for $\frac{\beta}{\alpha} = 0$, $a_n = \frac{(n-1)(n-2)}{\alpha n}$ and again for $n = 1$ $a_1 = 0$ and $y(x) = 1$ is the solution. Continuing in this fashion, we see that the series solution will terminate for β/α any positive integer as well as 0 and -1. For other values of β/α, we have $\frac{a_n}{a_{n-1}} = \frac{(n-\frac{\beta}{2}-1)(n-\frac{\beta}{\alpha}-2)}{\alpha n}$, which approaches ∞ as $n \to \infty$ and thus the ratio test yields a zero radius of convergence.

17b. Substituting $y = \sum\limits_{n=0}^{\infty} a_n x^{n+r}$ in the D.E. in standard form gives

$$\sum_{n=0}^{\infty} (n+r)(n+r-1) a_n x^{n+r} + \alpha \sum_{n=0}^{\infty} (n+r) a_n x^{n+r+1-s}$$

$$+ \beta \sum_{n=0}^{\infty} a_n x^{n+r+2-t} = 0.$$

If $s = 2$ and $t = 2$ the first term in each of the three series is $r(r-1)a_0 x^r$, $\alpha r a_0 x^{r-1}$, and $\beta a_0 x^r$, respectively. Thus we must have $\alpha r a_0 = 0$ which requires $r = 0$. Hence there is at most one solution of the assumed form.

17d. In order for the indicial equation to be quadratic in r it is necessary that the first term in the first series contribute to the indicial equation. This means that the first term in the second and the third series cannot appear before the first term of the first series. The first terms are $r(r-1)a_0 x^r$, $\alpha r a_0 x^{r+1-s}$, and $\beta a_0 x^{r+2-t}$, respectively. Thus if $s \leq 1$ and $t \leq 2$ the quadratic term will appear in the indicial equation.

Section 5.9, Page 271

1. It is clear that $x = 0$ is a singular point. The D.E. is in the standard form given in Theorem 5.8.1 with $xp(x) = 2$ and $x^2 q(x) = x$. Both are analytic at $x = 0$, so $x = 0$ is a regular singular point. Substituting

$$y = \sum_{n=0}^{\infty} a_n x^{n+r}$$ in the D.E., shifting indices

appropriately, and collecting coefficients of like powers of x yields

$$[r(r-1) + 2r]a_0 x^r + \sum_{n=1}^{\infty} [(r+n)(r+n+1)a_n + a_{n-1}]x^{r+n} = 0.$$

The indicial equation is $F(r) = r(r+1) = 0$ with roots $r_1 = 0$, $r_2 = -1$. Treating a_n as a function of r, we see that $a_n(r) = -a_{n-1}(r)/F(r+n)$, $n = 1,2,\ldots$ if $F(r+n) \neq 0$. Thus $a_1(r) = -a_0/F(r+1)$, $a_2(r) = a_0/F(r+1)F(r+2),\ldots$, and $a_n(r) = (-1)^n a_0/F(r+1)F(r+2)\ldots F(r+n)$, provided $F(r+n) \neq 0$ for $n = 1,2,\ldots$. For the case $r_1 = 0$, we have $a_n(0) = (-1)^n a_0/F(1)F(2) \ldots F(n) = (-1)^n a_0/n!(n+1)!$ so one solution is $y_1(x) = \sum\limits_{n=0}^{\infty} (-1)^n x^n / n!(n+1)!$ where we have

set $a_0 = 1$.

If we try to use the above recurrence relation for the case $r_2 = -1$ we find that $a_n(-1) = -a_{n-1}/n(n-1)$, which is undefined for $n = 1$. Thus we must follow the procedure described at the end of Section 5.8 to calculate a second solution of the form given in Eq.(6). Specifically, we use Eqs.(15) and (16) of that section to calculate a and $c_n(r_2)$ where $r_2 = -1$. Since

$r_1 - r_2 = 1 = N$, we have $a_N(r) = a_1(r) = -1/F(r+1)$. Hence

$a = \lim\limits_{r \to -1} [(r+1)(-1)/F(r+1)] = \lim\limits_{r \to -1} [-(r+1)/(r+1)(r+2)] = -1$.

Next

$$c_n(-1) = \frac{d}{dr}[(r+1)a_n(r)]\Big|_{r=-1} = (-1)^n \frac{d}{dr}[\frac{(r+1)}{F(r+1) \ldots F(r+n)}]\Big|_{r=-1},$$

where we have set $a_0 = 1$. Observe that $(r+1)/F(r+1) \ldots$

$F(r+n) = 1/[(r+2)^2(r+3)^2 \ldots (r+n)^2(r+n+1)] = 1/G_n(r)$.

Hence $c_n(-1) = (-1)^{n+1}G_n'(-1)/G_n^2(-1)$. Notice that

$G_n(-1) = 1^2 \cdot 2^2 \cdot 3^2 \ldots (n-1)^2 n = (n-1)!n!$ and

$G_n'(-1)/G_n(-1) = 2[1/1 + 1/2 + 1/3 + \ldots + 1/(n-1)] + 1/n =$

$H_n + H_{n-1}$. Thus $c_n(-1) = (-1)^{n+1}(H_n + H_{n-1})/(n-1)!n!$.

From Eq.(6) of Section 5.8 we obtain the second solution

$$y_2(x) = -y_1(x)\ln x + x^{-1}[1 - \sum\limits_{n=1}^{\infty} (-1)^n (H_n + H_{n-1})x^n/n!(n-1)!].$$

2. It is clear that $x = 0$ is a singular point. The D.E. is in the standard form given in Theorem 5.8.1 with $xp(x) = 3$ and $x^2q(x) = 1+x$. Both are analytic at $x = 0$, so $x = 0$ is a regular singular point. Substituting

$y = \sum\limits_{n=0}^{\infty} a_n x^{n+r}$ in the D.E., shifting indices

appropriately, and collecting coefficients of like powers of x yields

$$[r(r-1) + 3r + 1]a_0 x^r + \sum\limits_{n=1}^{\infty} \{[(r+n)(r+n+2) + 1]a_n$$

$$+ a_{n-1}\} x^{n+r} = 0.$$

The indicial equation is $F(r) = r^2 + 2r + 1 = (r+1)^2 = 0$ with the double root $r_1 = r_2 = -1$. Treating a_n as a function of r, we see that $a_n(r) = -a_{n-1}(r)/F(r+n)$, $n = 1, 2, \ldots$. Thus $a_1(r) = -a_0/F(r+1)$, $a_2(r) = a_0/F(r+1)F(r+2), \ldots$, and $a_n(r) = (-1)^n a_0/F(r+1)F(r+2)\ldots F(r+n)$. Setting $r = -1$ we find that $a_n(-1) = (-1)^n a_0/(n!)^2$, $n = 1, 2, \ldots$. Hence one solution is $y_1(x) = x^{-1}\sum\limits_{n=0}^{\infty}(-1)^n x^n/(n!)^2$ where we have set $a_0 = 1$. To find a second solution we follow the procedure described in Section 5.8 for the case when the roots of the indicial equation are equal. Specifically, the second solution will have the form given in Eq. (14) of that section. We must calculate $a_n'(-1)$. If we let $G_n(r) = F(r+1)\ldots F(r+n) = (r+2)^2(r+3)^2\ldots (r+n+1)^2$ and take $a_0 = 1$, then $a_n'(-1) = (-1)^n[1/G_n(r)]'$ evaluated $r = -1$. Hence $a_n'(-1) = (-1)^{n+1}G_n'(-1)/G_n^2(-1)$. But $G_n(-1) = (n!)^2$ and $G_n'(-1)/G_n(-1) = 2[1/1 + 1/2 + 1/3 + \ldots + 1/n] = 2H_n$. Thus a second solution is

$$y_2(x) = y_1(x)\ln x - 2x^{-1}\sum\limits_{n=1}^{\infty}(-1)^n H_n x^n/(n!)^2.$$

3. The roots of the indicial equation are $r_1 = r_2 = 0$ and thus the analysis is similar to that for Problem 2.

4. The roots of the indicial equation are $r_1 = -1$ and $r_2 = -2$ and thus the analysis is similar to that for Problem 1.

5. Since $x = 0$ is a regular singular point, substitute $y = \sum\limits_{n=0}^{\infty} a_n x^{n+r}$ in the D.E., shift indices appropriately, and collect coefficients of like powers of x to obtain $[r^2 - 9/4]a_0 x^r + [(r+1)^2 - 9/4]a_1 x^{r+1}$

$$+ \sum\limits_{n=2}^{\infty}\{[(r+n)^2 - 9/4]a_n + a_{n-2}\}\, x^{n+r} = 0.$$

The indicial equation is $F(r) = r^2 - 9/4 = 0$ with roots $r_1 = 3/2$, $r_2 = -3/2$. Treating a_n as a function of r we see that $a_n(r) = -a_{n-2}(r)/F(r+n)$, $n = 2,3,\ldots$ if $F(r+n) \neq 0$. For the case $r_1 = 3/2$, $F(r_1+1)$, which is the coefficient of x^{r_1+1}, is $\neq 0$ so we must set $a_1 = 0$. It follows that $a_3 = a_5 = \ldots = 0$. For the even coefficients, set $n = 2m$ so $a_{2m}(3/2) = -a_{2m-2}(3/2)/F(3/2 + 2m)$, $m = 1,2\ldots$. Thus $a_2(3/2) = -a_0/2^2 \cdot 1(1 + 3/2)$,

$a_4(3/2) = a_0/2^4 \cdot 2!(1 + 3/2)(2 + 3/2),\ldots$, and

$a_{2m}(3/2) = (-1)^m/2^{2m} m! \cdot (1 + 3/2)\ldots(m + 3/2)$. Hence one solution is

$$y_1(x) = x^{3/2}\left[1 + \sum_{m=1}^{\infty} \frac{(-1)^m}{m!(1 + 3/2)(2 + 3/2)\ldots(m + 3/2)} \left(\frac{x}{2}\right)^{2m}\right],$$

where we have set $a_0 = 1$. For this problem, the roots r_1 and r_2 of the indicial equation differ by an integer: $r_1 - r_2 = 3/2 - (-3/2) = 3$. Hence we can anticipate that there may be difficulty in calculating a second solution corresponding to $r = r_2$. This difficulty will occur in calculating $a_3(r) = -a_1(r)/F(r+3)$ because when $r = r_2 = -3/2$ we have $F(r_2+3) = F(r_1) = 0$. However, in this problem we are fortunate because $a_1 = 0$ and it will not be necessary to use the theory described at the end of Section 5.8. Notice for $r = r_2 = -3/2$ that the coefficient of x^{r_2+1} is $[(r_2+1)^2 - 9/4]a_1$, which does not vanish unless $a_1 = 0$. Thus the recurrence relation for the odd coefficients yields $a_5 = -a_3/F(7/2)$, $a_7 = -a_5/F(11/2) = a_3/F(11/2)F(7/2)$ and so forth. Substituting these terms into the assumed form we see that a multiple of $y_1(x)$ has been obtained and thus we may take $a_3 = 0$ without loss of generality. Hence $a_3 = a_5 = a_7 = \ldots = 0$. The even coefficients are given by $a_{2m}(-3/2) = -a_{2m-2}(-3/2)/F(2m - 3/2)$, $m = 1,2\ldots$.

Thus $a_2(-3/2) = -a_0/2^2 \cdot 1 \cdot (1 - 3/2)$,

$a_4(-3/2) = a_0/2^4 \cdot 2!(1 - 3/2)(2 - 3/2),\ldots$, and

$a_{2m}(-3/2) = (-1)^m a_0/2^{2m} m!(1 - 3/2)(2 - 3/2)\ldots(m - 3/2)$. Thus a second solution is

$$y_2(x) = x^{-3/2}[1 + \sum_{m=1}^{\infty} \frac{(-1)^m}{m!(1 - 3/2)(2 - 3/2) \cdots (m - 3/2)} (\frac{x}{2})^{2m}].$$

7. Apply the ratio test:
$$\lim_{m \to \infty} \frac{|(-1)^{m+1} x^{2m+2}/2^{2m+2}[(m+1)!]^2|}{|(-1)^m x^{2m}/2^{2m}(m!)^2|} = |x^2| \lim_{m \to \infty} \frac{1}{2^2(m+1)^2} = 0$$
for every x. Thus the series for $J_0(x)$ converges absolutely for all x.

12. If $\xi = \alpha x^{\beta}$, then $dy/dx = \frac{1}{2}x^{-1/2}f + x^{1/2}f'\alpha\beta x^{\beta-1}$ where f'

denotes $df/d\xi$. Find d^2y/dx^2 in a similar fashion and use algebra to show that f satisfies the D.E.
$$\xi^2 f'' + \xi f' + [\xi^2 - v^2]f = 0.$$

13. To compare $y'' - xy = 0$ with the D.E. of Problem 12, we must multiply by x^2 to get $x^2 y'' - x^3 y = 0$. Thus $2\beta = 3$, $\alpha^2\beta^2 = -1$ and $1/4 - v^2\beta^2 = 0$. Hence $\beta = 3/2$, $\alpha = 2i/3$ and $v = 1/3$ which yields the desired result.

14. First we verify that $J_0(\lambda_j x)$ satisfies the D.E. We know that $J_0(t)$ is a solution of the Bessel equation of order zero:
$$t^2 J_0''(t) + t J_0'(t) + t^2 J_0(t) = 0 \text{ or}$$
$$J_0''(t) + t^{-1} J_0'(t) + J_0(t) = 0.$$
Let $t = \lambda_j x$. Then
$$\frac{d}{dx} J_0(\lambda_j x) = \frac{d}{dt} J_0(t) \frac{dt}{dx} = \lambda_j J_0'(t)$$
$$\frac{d^2}{dx^2} J_0(\lambda_j x) = \lambda_j \frac{d}{dt}[J_0'(t)]\frac{dt}{dx} = \lambda_j^2 J_0''(t).$$
Substituting $y = J_0(\lambda_j x)$ in the given D.E. and making use of these results, we have
$$\lambda_j^2 J_0''(t) + (\lambda_j/t) \lambda_j J_0'(t) + \lambda_j^2 J_0(t) =$$
$$\lambda_j^2[J_0''(t) + t^{-1} J_0'(t) + J_0(t)] = 0.$$
Thus $y = J_0(\lambda_j x)$ is a solution of the given D.E. For the second part of the problem we follow the hint. First, rewrite the D.E. by multiplying by x to yield
$xy'' + y' + \lambda_j^2 xy = 0$, which can be written as
$(xy')' = -\lambda_j^2 xy$. Now let $y_i(x) = J_0(\lambda_i x)$ and

$y_j(x) = J_0(\lambda_j x)$ and we have, respectively:

$$(xy_i')' = -\lambda_i^2 xy_i$$

$$(xy_j')' = -\lambda_j^2 xy_j.$$

Now multiply the first equation by y_j, the second by y_i, integrate each from 0 to 1, and subtract the second from the first:

$$\int_0^1 [y_j(xy_i')' - y_i(xy_j')']dx = -(\lambda_i^2 - \lambda_j^2) \int_0^1 xy_i y_j dx.$$

If we integrate each term on the left side once by parts and note that $y_i = y_j = 0$ at $x = 0$ and $x = 1$, we find that the left side of this equation is identically zero. Hence the right side is identically zero and for $\lambda_i \neq \lambda_j$ this gives the desired result.

CHAPTER 6

1. The graph of f(t) is shown.
    Since the function is
    continuous on each interval,
    but has a jump discontinuity
    at t = 1, f(t) is piecewise
    continuous.

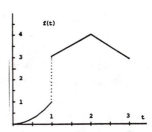

5b. Since t^2 is continuous for $0 \leq t \leq A$ for any positive A
    and since $t^2 \leq e^{at}$ for any a > 0 and for t sufficiently
    large, it follows from Theorem 6.1.2 that $\mathcal{L}\{t^2\}$ exists
    for s > 0. $\mathcal{L}\{t^2\} = \int_0^\infty e^{-st}t^2 dt = \lim_{M \to \infty} \int_0^M e^{-st}t^2 dt$

    $$= \lim_{M \to \infty} [\frac{-t^2}{s}e^{-st}\big|_0^M + \frac{2}{s}\int_0^M e^{-st}t\, dt]$$

    $$= \frac{2}{s} \lim_{M \to \infty} [-\frac{1}{s}te^{-st}\big|_0^M + \frac{1}{s}\int_0^M e^{-st}dt]$$

    $$= \frac{2}{s^2} \lim_{M \to \infty} - \frac{1}{s}e^{-st}\big|_0^M = \frac{2}{s^3}.$$

6. That f(t) = cosat satisfies the hypotheses of Theorem
    6.1.2 can be verified by recalling that $|cosat| \leq 1$ for
    all t. To determine $\mathcal{L}\{cosat\} = \int_0^\infty e^{-st}$ cosatdt we

    must integrate by parts twice to get $\int_0^\infty e^{-st}$ cosatdt =

    $\lim_{M \to \infty} [(-s^{-1}e^{-st}cosat + as^{-2}e^{-st}sinat)\big|_0^M - (a^2/s^2)$

    $\int_0^M e^{-st}cosatdt]$. Evaluating the first two terms, letting
    $M \to \infty$, and adding the third term to both sides, we
    obtain
    $[1 + a^2/s^2]\int_0^\infty e^{-st}cosatdt = 1/s$, s > 0. Division by

    $[1 + a^2/s^2]$ and simplification yields the desired
    solution.

9. From the definition for coshbt we have
    $\mathcal{L}\{e^{at}coshbt\} = \mathcal{L}\{\frac{1}{2}[e^{(a+b)t} + e^{(a-b)t}]\}$. Using the

    linearity property of $\mathcal{L}$, Eq.(5), the right side becomes
    $\frac{1}{2}\mathcal{L}\{e^{(a+b)t}\} + \frac{1}{2}\mathcal{L}\{e^{(a-b)t}\}$ which can be evaluated using

    the result of Example 5 and thus

$$\mathcal{L}\{e^{at}\cosh bt\} = \frac{1/2}{s-(a+b)} + \frac{1/2}{s-(a-b)}$$

$$= \frac{s-a}{(s-a)^2-b^2}, \quad \text{for } s-a > |b|.$$

13. We write $\sin at = (e^{iat} - e^{-iat})/2i$, then the linearity of the Laplace transform operator allows us to write
$\mathcal{L}\{e^{at}\sin bt\} = (1/2i)\mathcal{L}\{e^{(a+ib)t}\} - (1/2i)\mathcal{L}\{e^{(a-ib)t}\}$.
Each of these two terms can be evaluated by using the result of Example 5, where we now have to require s to be greater than the real part of the complex numbers a + ib and
a - ib in order for the integrals to converge. Complex algebra then gives the desired result. An alternate method of evaluation would be to use integration on the integral appearing in the definition of $\mathcal{L}\{e^{at}\sin bt\}$, but that method is very cumbersome.

16. As in Problem 13, $\mathcal{L}\{t\sin at\} = (1/2i)\mathcal{L}\{te^{iat}\} -$
$(1/2i)\mathcal{L}\{te^{-iat}\}$. Using the result of Problem 15 we obtain $\mathcal{L}\{t\sin at\} = (1/2i)[(s-b)^{-2} - (s+b)^{-2}]$ where b = ia and s > 0. Hence $\mathcal{L}\{t\sin at\} = 2as/(s^2+a^2)^2$, s > 0.

19. Use the approach shown in Problem 16, with a second integration by parts.

21. The integral $\int_0^A (t^2 + 1)^{-1}dt$ can be evaluated in terms of the arctan function and then definition (3) can be used. To illustrate Theorem 6.1.1, however, consider that
$\frac{1}{t^2+1} < \frac{1}{t^2}$ for $t \geq 1$ and from Example 3, $\int_1^\infty t^{-2}dt$
converges and hence $\int_1^\infty (t^2 + 1)^{-1}dt$ also converges.
$\int_0^1 (t^2 + 1)^{-1}dt$ is finite and hence does not affect the convergence of $\int_0^\infty (t^2 + 1)^{-1}dt$ at infinity.

25. If we let u = f and $dv = e^{-st}dt$ then $F(s) = \int_0^\infty e^{-st}f(t)dt$
$= \lim_{M \to \infty} -\frac{1}{s}e^{-st}f(t)\big|_0^M + \frac{1}{s}\int_0^\infty e^{-st}f'(t)dt$. Now use an argument similar to that given to establish Theorem 6.1.2.

27a. Make a transformation of variables with x = st and
     dx = sdt. Then use the definition of $\Gamma(p+1)$ from
     Problem 26.

27d. Use the definition of $\mathcal{L}\{t^{1/2}\}$ and integrate by parts once
     to get $\mathcal{L}\{t^{1/2}\} = (1/2s)\mathcal{L}\{t^{-1/2}\}$. The result follows from
     part (c).

Section 6.2, Page 289

Problems 1 through 10 are solved by using partial fractions
and algebra to manipulate the given function into a form
matching one of the functions appearing in the middle column
of Table 6.2.1.

2.   We have $\dfrac{4}{(s-1)^3} = 2\dfrac{2!}{(s-1)^{2+1}}$ and thus the inverse Laplace
     transform is $2t^2e^t$, using line 11.

4.   We have $\dfrac{3s}{s^2-s-6} = \dfrac{3s}{(s-3)(s+2)} = \dfrac{9/5}{s-3} + \dfrac{6/5}{s+2}$ using partial
     fractions. Thus $(9/5)e^{3t} + (6/5)e^{-2t}$ is the inverse
     transform, from line 2.

7.   We have $\dfrac{2s+1}{s^2-2s+2} = \dfrac{2s+1}{(s-1)^2+1} = \dfrac{2(s-1)}{(s-1)^2+1} + \dfrac{3}{(s-1)^2+1}$, where
     we first used the concept of completing the square (in
     the denominator) and then added and subtracted
     appropriately to put the numerator in the desired form.
     Lines 9 and 10 may now be used to find the desired
     result.

In each of the Problems 11 through 23 it is assumed that the
I.V.P. has a solution $y = \phi(t)$ which, with its first two
derivatives, satisfies the conditions of the Corollary to
Theorem 6.2.1.

11. Take the Laplace transform of the D.E. to get
     $s^2Y(s) - sy(0) - y'(0) - [sY(s) - y(0)] - 6Y(s) = 0$.
     Using the I.C. and solving for Y(s) we obtain
     $Y(s) = \dfrac{s-2}{s^2-s-6}$. Following the pattern of Eq.(12) we have
     $\dfrac{s-2}{s^2-s-6} = \dfrac{a}{s+2} + \dfrac{b}{s-3} = \dfrac{a(s-3)+b(s+2)}{(s+2)(s-3)}$. Equating like

powers in the numerators we find a+b = 1 and −3a+2b = −2.
Thus a = 4/5 and b = 1/5 and $Y(s) = \dfrac{4/5}{s+2} + \dfrac{1/5}{s-3}$, which
yields the desired solution using Table 6.2.1.

14. Taking the Laplace transform we have $s^2Y(s) - sy(0) -$
    $y'(0) - 4[sY(s)-y(0)] + 4Y(s) = 0$. Using the I.C. and
    solving for Y(s) we find $Y(s) = \dfrac{s-3}{s^2-4s+4}$. Since the
    denominator is a perfect square, the partial fraction
    form is $\dfrac{s-3}{s^2-4s+4} = \dfrac{a}{(s-2)^2} + \dfrac{b}{s-2}$. Solving for a and b,
    as shown in examples of this section or in Problem 11, we
    find a = −1 and b = 1. Thus $Y(s) = \dfrac{1}{s-2} - \dfrac{1}{(s-2)^2}$, from
    which we find $y(t) = e^{2t} - te^{2t}$.

15. Note that $Y(s) = \dfrac{2s-4}{s^2-2s-2} = \dfrac{2s-4}{(s-1)^2-3} = \dfrac{2(s-1)}{(s-1)^2-3} - \dfrac{2}{(s-1)^2-3}$.
    Three formulas in Table 6.2.1 are now needed: F(s−c) in
    line 14 in conjunction with the ones for coshat and
    sinhat, lines 7 and 8.

17. The Laplace transform of the D.E. is
    $s^4Y(s) - s^3y(0) - s^2y'(0) - sy''(0) - y'''(0) - 4[s^3Y(s)-s^2y(0)$
    $-sy'(0) - y''(0)] + 6[s^2Y(s) - sy(0) - y'(0)] - 4[sY(s) - y(0)]$
    $+ Y(s) = 0$. Using the I.C. and solving for Y(s) we find
    $Y(s) = \dfrac{s^2 - 4s + 7}{s^4-4s^3+6s^2-4s+1}$. The correct partial fraction
    form for this is $\dfrac{1}{(s-1)^4} + \dfrac{b}{(s-1)^3} + \dfrac{c}{(s-1)^2} + \dfrac{d}{s-1}$.
    Setting this equal to Y(s) above and equating the
    numerators we have $s^2-4s+7 = a + b(s-1) + c(s-1)^2 +$
    $d(s-1)^3$. Solving for a,b,c, and d and use of Table 6.2.1
    yields the desired solution.

20. The Laplace transform of the D.E. is
    $s^2Y(s) - sy(0) - y'(0) + \omega^2Y(s) = s/(s^2+4)$. Applying the
    I.C. and solving for Y(s) we get $Y(s) = s/[(s^2+4)(s^2+\omega^2)]$
    $+ s/(s^2+\omega^2)$. Decomposing the first term by partial
    fractions we have

$$Y(s) = \frac{s}{(\omega^2-4)(s^2+4)} - \frac{s}{(\omega^2-4)(s^2+\omega^2)} + \frac{s}{s^2+\omega^2}$$

$$= (\omega^2-4)^{-1}[\frac{(\omega^2-5)s}{s^2+\omega^2} + \frac{s}{s^2+4}].$$

Then, using Table 6.2.1, we have

$$y = (\omega^2-4)^{-1}[(\omega^2-5)\cos\omega t + \cos 2t].$$

22. Solving for $Y(s)$ we find $Y(s) = 1/[(s-1)^2 + 1] +$
    $1/(s+1)[(s-1)^2 + 1]$. Using partial fractions on the
    second term we obtain

$$Y(s) = 1/[(s-1)^2 + 1] + \{1/(s+1) - (s-3)/[(s-1)^2 + 1]\}/5$$

$$= (1/5)\{(s+1)^{-1} - (s-1)[(s-1)^2 + 1]^{-1} + 7[(s-1)^2 + 1]^{-1}\}.$$

Hence, $y = (1/5)(e^{-t} - e^t\cos t + 7e^t\sin t)$.

24. Under the standard assumptions, the Lapace transform of
    the left side of the D.E. is $s^2 Y(s) - sy(0) - y'(0) +$
    $4Y(s)$. To transform the right side we must revert to the
    definition of the Laplace trasnform to determine
    $\int_0^\infty e^{-st}f(t)\,dt$. Since $f(t)$ is piecewise continuous we are

    able to calculate $\mathcal{L}\{f(t)\}$ by

$$\int_0^\infty e^{-st}f(t)\,dt = \int_0^\pi e^{-st}\,dt + \lim_{M\to\infty}\int_\pi^M (e^{-st})(0)\,dt$$

$$= \int_0^\pi e^{-st}\,dt = (1 - e^{-\pi s})/s.$$

Hence, the Laplace transform $Y(s)$ of the solution is
given by $Y(s) = s/(s^2+4) + (1 - e^{-\pi s})/s(s^2+4)$.

27b. The Taylor series for f about t = 0 is

$$f(t) = \sum_{n=0}^\infty (-1)^n t^{2n}/(2n+1)!, \text{ which is obtained from}$$

part (a) by dividing each term of the sine series by t.
Also, f is continuous for t > 0 since $\lim_{t\to 0+}(\sin t)/t = 1$.

Assuming that we can compute the Laplace transform of f

term by term, we obtain $\mathcal{L}\{f(t)\} = \mathcal{L}\{\sum_{n=0}^\infty (-1)^n t^{2n}/(2n+1)!\}$

$$= \sum_{n=0}^\infty [(-1)^n/(2n+1)!\,\mathcal{L}\{t^{2n}\}$$

$$= \sum_{n=0}^{\infty} [(-1)^n (2n)!/(2n+1)!] s^{-(2n+1)}$$

$$= \sum_{n=0}^{\infty} [(-1)^n/(2n+1)] s^{-(2n+1)}, \text{ which converges for } s > 1.$$

The Taylor series for arctanx is given by

$$\sum_{n=0}^{\infty} (-1)^n x^{2n+1}/(2n+1).$$ Comparing $\mathcal{L}\{f(t)\}$ with the Taylor

series for arctanx, we conclude that
$\mathcal{L}\{f(t)\} = \arctan(1/s), \ s > 1.$

30. Setting n = 2 in Problem 28b, we have

$$\mathcal{L}\{t^2 \sin bt\} = \frac{d^2}{ds^2}[b/(s^2+b^2)] = \frac{d}{ds}[-2bs/(s^2+b^2)^2] =$$

$$-2b/(s^2+b^2)^2 + 8bs^2/(s^2+b^2)^3 = 2b(3s^2-b^2)/(s^2+b^2)^3.$$

32. Using the result of Problem 28a. we have

$$\mathcal{L}\{te^{at}\} = -\frac{d}{ds}(s-a)^{-1} = (s-a)^{-2}$$

$$\mathcal{L}\{t^2 e^{at}\} = -\frac{d}{ds}(s-a)^{-2} = 2(s-a)^{-3}.$$

$$\mathcal{L}\{t^3 e^{at}\} = -\frac{d}{ds}2(s-a)^{-3} = 3!(s-a)^{-4}.$$ Continuing in this

fashion or using induction we obtain the desired result.

36a. Taking the Laplace transform of the D.E. we obtain

$$\mathcal{L}\{y''\} + \mathcal{L}\{ty\} = \mathcal{L}\{y''\} - \mathcal{L}\{-ty\}$$
$$= s^2 Y(s) - sy(0) - y'(0) - Y'(s) = 0.$$

Hence, Y satisfies $Y' - s^2 Y = s.$

38a. Follow the hint and apply L'Hopital's rule recalling that
Q(s) has distinct zeros.

Section 6.3, Page 296

2. From the definition of $u_c(t)$ we have

$$g(t) = (t-3)u_2(t) - (t-2)u_3(t) = \begin{cases} 0 - 0 = 0, & 0 \le t < 2 \\ (t-3) - 0 = t-3, & 2 \le t < 3. \\ (t-3) - (t-2) = -1, & 3 \le t \end{cases}$$

The graph of y = g(t) is:

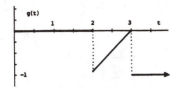

4. As indicated in the discussion following Example 1, the
    unit step function can be used to translate a given
    function f, with domain t≥0, a distance c to the right by
    the multiplication $u_c(t)f(t-c)$. Hence the required graph
    of $y = f(t-3)u_3(t)$ for $f(t) = \sin t$ is:

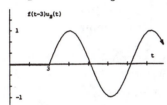

8. In order to use Theorem 6.3.1 we must write f(t) in terms
    of $u_c(t)$. Since $t^2 - 2t + 2 = (t-1)^2 + 1$ (by completing
    the square), we can thus write $f(t) = u_1(t)g(t-1)$, where
    $g(t) = t^2+1$. Now applying Theorem 6.3.1 we have
    $$\mathcal{L}\{f(t)\} = \mathcal{L}\{u_1(t)g(t-1)\} = e^{-s}\mathcal{L}\{g(t)\} = e^{-s}(2/s^3 + 1/s).$$

14. Use partial fractions to write
    $F(s) = e^{-2s}[(s-1)^{-1} - (s+2)^{-1}]/3$. For ease in
    calculations let us define $G(s) = (s-1)^{-1}$ and $H(s) =$
    $(s+2)^{-1}$. Then
    $F(s) = [e^{-2s}G(s) - e^{-2s}H(s)]/3$. Using the fact that
    $\mathcal{L}\{e^{at}\} = (s-a)^{-1}$ and applying Theorem 6.3.1, we have
    $F(s) = [e^{-2s}\mathcal{L}\{e^t\} - e^{-2s}\mathcal{L}\{e^{-2t}\}]/3$. Thus,
    $F(s) = [\mathcal{L}\{u_2(t)e^{(t-2)}\} - \mathcal{L}\{u_2(t)e^{-2(t-2)}\}]/3$. Using the
    linearity of the Laplace transform, we have
    $\mathcal{L}\{f(t)\} = \mathcal{L}\{u_2(t)[e^{t-2} - e^{-2(t-2)}]/3\}$. Hence,
    $f(t) = [u_2(t)(e^{t-2} - e^{-2(t-2)})]/3$. An alternate method is
    to complete the square in the denominator:
    $$F(s) = \frac{e^{-2s}}{(s+1/2)^2 - 9/4}. \quad \text{This gives}$$
    $f(t) = (2/3)u_2(t)e^{-(t-2)/2}\sinh\frac{3}{2}(t-2)$, which is the same

as that found above.

21. By completing the square in the denominator of F we can
    write $F(s) = (2s+1)/[(2s+1)^2 + 4]$. This has the form
    $G(2s+1)$ where $G(u) = u/(u^2+4)$. We must find
    $\mathcal{L}^{-1}\{G(2s+1)\}$. Applying the results of Problem 19(c), we
    have

    $$\mathcal{L}^{-1}\{F(s)\} = \frac{1}{2}e^{-t/2}\cos\left(\frac{2t}{2}\right).$$

22. If the approach of Problem 21 is used we find
    $f(t) = (1/3)e^{2t/3}\sinh(t/3)$, which is equivalent to the
    given answer using the definition of sinht.

27. Assuming that term-by-term integration of the infinite
    series is permissible and recalling that
    $\mathcal{L}\{u_c(t)\} = e^{-cs}/s$ for $s > 0$, we have

    $$\mathcal{L}\{f(t)\} = (1/s) + \sum_{k=1}^{\infty} (-1)^k \mathcal{L}\{u_k(t)\}$$

    $$= (1/s) + \sum_{k=1}^{\infty}(-1)^k e^{-ks}/s = [\sum_{k=0}^{\infty}(-e^{-s})^k]/s. \quad \text{We recognize}$$

    the last infinite series as the geometric series, $\sum_{k=0}^{\infty} ar^k$,

    with $r = -e^{-s}$. This series converges to $[1/(1+e^{-s})]$ if
    $|r| < 1$. Hence, $\mathcal{L}\{f(t)\} = (1/s)[1/(1+e^{-s})]$, $s > 0$.

28. Using the definition of the Laplace transform we have
    $F(s) = \mathcal{L}\{f(t)\} = \int_0^\infty e^{-st}f(t)\,dt$. Since f is periodic with
    period T, we have $f(t+T) = f(t)$. This suggests that we
    rewrite the improper integral as $\int_0^\infty e^{-st}f(t)\,dt =$

    $$\sum_{n=0}^{\infty}\int_{nT}^{(n+1)T}e^{-st}f(t)\,dt. \quad \text{The periodicity of f also suggests}$$

    that we make the change of variable $t = r + nT$. Hence,

    $$F(s) = \sum_{n=0}^{\infty}\int_0^T e^{-s(r+nT)}f(r+nT)\,dr = \sum_{n=0}^{\infty}(e^{-sT})^n\int_0^T e^{-rs}f(r)\,dr,$$

    where we have used the fact that
    $f(r+nT) = f(r+(n-1)T) = \ldots = f(r+T) = f(r)$ from the

definition that f is periodic. We recognize this last
series as the geometric series, $\sum_{n=0}^{\infty} au^n$, with a =
$\int_0^T e^{-rs}f(r)dr$ and u = e^{-sT}. The geometric series converges
to a/(1-u) for |u| < 1 and consequently we obtain
$F(s) = (1 - e^{-sT})^{-1} \int_0^T e^{-rs}f(r)dr$, s > 0.

30. The function f is periodic with period 2. The result of
Problem 28 gives us $\mathcal{L}\{f(t)\} = \int_0^2 e^{-st}f(t)dt/(1-e^{-2s})$.
Calculating the integral we have
$$\int_0^2 e^{-st}f(t)dt = \int_0^1 e^{-st}dt - \int_1^2 e^{-st}dt$$
$$= (1-e^{-s})/s + (e^{-2s}-e^{-s})/2s$$
$$= (e^{-2s}-2e^{-s}+1)/s$$
$$= (1-e^{-s})^2/s.$$ Since the denominator of
$\mathcal{L}\{f(t)\}$, $1 - e^{-2s}$, may be written as $(1-e^{-s})(1+e^{-s})$ we
obtain the desired answer.

Section 6.4, Page 303

1. f(t) can be written in the form f(t) = 1 - $u_{\pi/2}(t)$ and
thus the Laplace transform of the D.E. is
$(s^2+1)Y(s) - sy(0) - y'(0) = (1/s) - e^{-\pi s/2}/s.$
Introducing the I.C. and solving for Y(s), we obtain
$Y(s) = (s^2+1)^{-1} + [s(s^2+1)]^{-1} - e^{-\pi s/2}/s(s^2+1).$ Using
partial fractions on the second and third term we find
$Y(s) = (1/s) + (s^2+1)^{-1} - s/(s^2+1) - e^{-\pi s/2}/s + e^{-\pi s/2}s/(s^2+1).$
The inverse transform of the first three terms can be
obtained directly from Table 6.2.1. Using Theorem 6.3.1
to find the inverse transform of the last two terms we
have
$\mathcal{L}^{-1}\{e^{-s/2}/s\} = u_{\pi/2}(t)g(t - \pi/2)$ where $g(t) = \mathcal{L}^{-1}\{1/s\} = 1$
and $\mathcal{L}^{-1}\{e^{-s/2}s/(s^2+1)\} = u_{\pi/2}(t)h(t - \pi/2)$ where
$h(t) = \mathcal{L}^{-1}\{s/(s^2+1)\} = \cos t.$ Hence,
y = 1 + sin t - cos t + $u_{\pi/2}(t)[\cos(t - \pi/2) - 1]$
  = 1 + sin t - cos t - $u_{\pi/2}(t)[1 - \sin t].$

3. According to Theorem 6.3.1,
$\mathcal{L}\{u_{2\pi}(t)\sin(t-2\pi)\} = e^{-2\pi s}\mathcal{L}\{\sin t\} = e^{-2\pi s}/(s^2+1).$
Transforming the D.E., we have

$(s^2+4)Y(s) - sy(0) - y'(0) = 1/(s^2+1) - e^{-2\pi s}/(s^2+1)$.
Introducing the I.C. and solving for Y(s), we obtain
$Y(s) = (1-e^{-2\pi s})/(s^2+1)(s^2+4)$. We apply partial
fractions to write
$Y(s) = [s^2+1)^{-1} - (s^2+4)^{-1} - e^{-2\pi s}(s^2+1)^{-1} + e^{-2\pi s}(s^2+4)^{-1}]/3$.
We compute the inverse transform of the first two terms
directly from Table 6.2.1 after noting that
$(s^2+4)^{-1} = (1/2)[2/(s^2+4)]$. We apply Theorem 6.3.1 to the
last two terms to obtain the solution,
$y = (1/3)\{\sin t - (1/2)\sin 2t - u_{2\pi}(t)[\sin(t-2\pi) -$
$(1/2)\sin 2(t-2\pi)]\}$. This may be simplified using
trigonometric identities to
$y = [(2\sin t - \sin 2t)(1-u_{2\pi}(t))]/6$.

8. Taking the Laplace transform, applying the I.C. and using
    Theorem 6.3.1 (or referring to Example 2) we have
    $(s^2+s+5)/4)Y(s) = (1-e^{-\pi s/2})/s^2$. Thus $Y(s) =$

    $$\frac{1-e^{-s/2}}{s^2(s^2+s+5/4)}$$

    $= (1-e^{-\pi s/2})\left\{\dfrac{4/5}{s^2} - \dfrac{16/25}{s} + \dfrac{(16/25)s-4/25}{(s+1/2)^2+1}\right\} = (1-e^{-\pi s/2}$

)H(s), where we have used partial fractions and completed
    the square in the denominator of the last term. Since
    the numerator of the last term of H can be written as
    $\dfrac{16}{25}[(s+1)/2) - 3/4]$, we see that

    $\mathcal{L}^{-1}\{H(s)\} = (4/25)(5t - 4 + 4e^{-t/2}\cos t - 3e^{-t/2}\sin t)$,
    which yields the desired solution.

10. Note that $g(t) = \sin t - u_\pi(t)\sin t = \sin t + u_\pi(t)\sin(t-\pi)$.
    Proceeding as in Problem 8 we find
    $Y(s) = (1+e^{-\pi s})\dfrac{1}{(s^2+1)(s^2+s+5/4)}$. The correct partial

    fraction expansion of the quotient is $\dfrac{as+b}{s^2+1} + \dfrac{cs+d}{s^2+s+5/4}$,
    where $a+c = 0$, $a+b+d = 0$, $(5/4)a+b+c = 0$ and
    $(5/4)b+d = 1$ by equating coefficients. Solving for the
    constants yields the desired solution.

14. Since f is periodic with period 2π, we can apply the
    result of Problem 28 in Section 6.3 to obtain
    $\mathcal{L}\{f(t)\} = \int_0^{2\pi} e^{-st}f(t)dt/(1 - e^{-2\pi s}) = 1/s(1 + e^{-\pi s})$.
    Thus, the transformed equation is

$(s^2+1)Y(s) - sy(0) - y'(0) = [s(1 + e^{-\pi s})]^{-1}$. Introducing the I.C. and solving gives

$$Y(s) = \frac{s}{s^2+1} + \frac{1}{(s^2+1)s(1+e^{-s})} = \frac{s}{s^2+1} + \frac{1}{(1+e^{-s})}[\frac{1}{s} - $$

$\frac{s}{s^2+1}]$ where partial fractions have been used to get the terms in square brackets. The inverse transform of this, however, cannot be found in Table 6.2.1. Recall that Problem 27, Section 6.3, gave a result that had $(1 + e^{-s})$ in the denominator. This indicates that we should attempt to write the second term in the expression for Y(s) as a series. In this case we can write $[1 + e^{-\pi s}]^{-1}$ as the sum of the geometric series,

$$[1 + e^{-\pi s}]^{-1} = \sum_{n=0}^{\infty} (-e^{-\pi s})^n = 1 + \sum_{n=1}^{\infty} (-e^{-\pi s})^n. \quad \text{Thus}$$

$$Y(s) = s/(s^2+1) + [1 + \sum_{n=1}^{\infty} (-e^{-\pi s})^n][(1/s) - s/(s^2+1)]$$

$$= (1/s) + \sum_{n=1}^{\infty} (-1)^n e^{-n\pi s} [(1/s) - s/(s^2+1)]. \quad \text{Assuming}$$

that term-by-term inversion of the infinite series is permissible, we apply Theorem 6.3.1 and Table 6.2.1 to

obtain $y = a + \sum_{n=1}^{\infty} (1)^n u_{n\pi}(t) [1 - \cos(t-n\pi)]$.

Section 6.5, Page 307

1. Proceeding as in the Example, we take the Laplace transform of the D.E. and apply the I.C.:
    $(s^2 + 2s + 2)Y(s) = s + 2 + e^{-\pi s}$. Thus,
    $Y(s) = (s+2)/[(s+1)^2 + 1] + e^{-\pi s}/[(s+1)^2 + 1]$. We write
    the first term as $(s+1)/[(s+1)^2 + 1] + 1/[(s+1)^2 + 1]$.
    Applying Theorem 6.3.1 and using Table 6.2.1, we obtain
    the solution,
    $y = e^{-t}\cos t + e^{-t}\sin t - u_\pi(t)e^{(t-\pi)}\sin t$. Note that
    $\sin(t-\pi) = -\sin t$.

3. Taking the Laplace transform of the D.E. and applying the
    I.C., we obtain $(s^2 + 2s + 1)Y(s) = 2 + e^{-2\pi s}/s$ where
    $\mathcal{L}\{\delta(t)\} = 1$. Solving for Y(s) and using partial

fractions on the last term, we have

$Y(s) = 2/(s+1)^2 + e^{-2\pi s}[(1/s) - 1/(s+1) - 1/(s+1)^2]$.   The

inverse transform of $2/(s+1)^2$ can be found in Table 6.2.1
or by using the result of Problem 28 of Section 6.2.
Hence, by Theorem 6.3.1, we have

$y = \mathcal{L}^{-1}\{Y(s)\} = 2te^{-t} + u_{2\pi}(t)[1 - e^{-(t-2\pi)} - (t-2\pi)e^{-(t-2\pi)}]$.

5.   Following the procedure of earlier problems we have

$Y(s) = (s^2+2)/[(s^2+1)(s^2+2s+3)] + e^{-\pi s}/(s^2+2s+3)$.   The

partial fraction expansion of the first term is
$(as+b)/(s^2+1) + (cs+d)/(s^2+2s+3)$, where $a + c = 0$,
$2a + b + d = 1$, $3a + 2b + c = 0$ and $3b + d = 2$.

7.   Taking the Laplace transform of the D.E. yields

$(s^2+1)Y(s) - y'(0) = \int_0^t e^{-st}\delta(t-\pi)\cos t\,dt$.   Since $\delta(t-\pi) = 0$

for $t \neq \pi$ the integral on the right is equal to

$\int_{-\infty}^{\infty} e^{-st}\delta(t-\pi)\cos t\,dt$ which equals $e^{-\pi s}\cos\pi$ from Eq.(16).

Substituting for $y'(0)$ and solving for $Y(s)$ gives the
desired solution.

13b. Substituting for $f(t)$ we have

$y = \int_0^t e^{-(t-\tau)}\delta(\tau-\pi)\sin(t-\tau)\,d\tau$.   We know that the

integration variable is always less than t (the upper
limit) and thus for $t < \pi$ we have $\tau < \pi$ and thus
$\delta(\tau-\pi) = 0$.   Hence $y = 0$ for $t < \pi$.   For $t > \pi$ utilize
Eq.(16).

Section 6.6, Page 313

1c. Using the format of Eqs.(2) and (3) we have

$$f*(g*h) = \int_0^t f(t-\tau)(g*h)(\tau)\,d\tau$$

$$= \int_0^t f(t-\tau)[\int_0^t g(\tau-\eta)h(\eta)\,d\eta]\,d\tau$$

$$= \int_0^t [\int_\eta^t f(t-\tau)g(\tau-\eta)\,d\tau]h(\eta)\,(d\eta).$$

The last double integral is obtained from the previous
line by interchanging the order of the η and τ
integration. Making the change of variable $\omega = \tau - \eta$ on
the inside integral and comparing the result with $(f*g)*h$
yields the desired result.

4.   It is possible to determine $f(t)$ explicitly by using

integration by parts and then to find its transform $F(s)$.
However, it is much more convenient to apply Theorem
6.6.1. Let us define $g(t) = t^2$ and $h(t) = \cos 2t$. Then,
$f(t) = \int_0^t g(t-\tau) h(\tau) d\tau$. Using Table 6.2.1, we have

$G(s) = \mathcal{L}\{g(t)\} = 2/s^3$ and $H(s) = \mathcal{L}\{h(t)\} = s/(s^2+4)$.
Hence, by Theorem 6.6.1,
$\mathcal{L}\{f(t)\} = F(s) = G(s)H(s) = 2/s^2(s^2+4)$.

8. As was done in Example 1 think of $F(s)$ as the product of
s^{-4} and $(s^2+1)^{-1}$ which, according to Table 6.2.1, are the
transforms of $t^3/6$ and $\sin t$, respectively. Hence, by
Theorem 6.6.1, the inverse transform of $F(s)$ is
$f(t) = (1/6)\int_0^t (t-\tau)^3 \sin\tau d\tau$.

13. We take the Laplace transform of the D.E. and apply the
I.C.: $(s^2 + 2s + 2)Y(s) = \alpha/(s^2 + \alpha^2)$. Solving for $Y(s)$,
we have $Y(s) = [\alpha/(s^2+\alpha^2)][(s+1)^2 + 1]^{-1}$, where the second
factor has been written in a convenient way by completing
the square. Thus $Y(s)$ is seen to be the product of the
transforms of $\sin\alpha t$ and $e^{-t}\sin t$ respectively. Hence,
according to Theorem 6.6.1, $y = \int_0^t e^{-(t-\tau)}\sin(t-\tau)\sin\alpha\tau d\tau$.

15. Proceeding as in the above problems we obtain
$$Y(s) = \frac{s}{s^2+s+5/4} + \frac{1-e^{-s}}{s(s^2+s+5/4)}$$
$$= \frac{(s+1/2) - 1/2}{(s+1/2)^2+1} + \frac{1-e^{-s}}{s} \cdot \frac{1}{(s+1/2)^2+1},$$
where the first term is obtained by completing the square
in the denominator and the second term is written as the
product of two terms whose inverse transforms are known
so that Theorem 6.6.1 can be used. Note that
$\mathcal{L}^{-1}\{(1-e^{-s})/s\} = 1 - u_\pi(t)$. Also note that a different
form of the same solution would be obtained by writing
the second term as $(1-e^{-\pi s})(\frac{a}{s} + \frac{bs + c}{(s+1/2)^2+1})$ and solving
for a, b and c. In this case $\mathcal{L}^{-1}\{1-e^{-\pi s}\} = \delta(t) - \delta(t-\pi)$
from Section 6.5.

17. Taking the Laplace transform, using the I.C. and solving
we have $Y(s) = (s+3)/(s+1)(s+2) + s/(s^2+\alpha^2)(s+1)(s+2)$.

As in Problem 15, there are several correct ways the second term can be treated in order to use the convolution integral. In order to obtain the desired answer, write the second term as

$$\frac{s}{s^2+\alpha^2}\left(\frac{a}{s+1} + \frac{b}{s+2}\right)$$ and solve for a and b.

20. To find $\Phi(s)$ you must recognize the integral that appears in the equation as a convolution integral.

CHAPTER 7

Section 7.1, Page 322

5. Let $x_1 = u$ and $x_2 = u'$; then $x_1' = x_2$ is the first of the
    desired pair of equations. The second equation is
    obtained by substituting $u'' = x_2'$, $u' = x_2$, and $u = x_1$ in
    the given D.E. The I.C. becomes $x_1(0) = u_0$, $x_2(0) = u_0'$.

8. We follow the steps outlined in Problem 7. Solve the
    first D.E. for x_2 to obtain $x_2 = \frac{3}{2}x_1 - \frac{1}{2}x_1'$. Substitute
    this into the second D.E. to obtain $x_1'' - x_1' - 2x_1 = 0$,
    which has the solution $x_1 = c_1 e^{2t} + c_2 e^{-t}$. Differentiating
    this and substituting into the above equation for x_2
    yields $x_2 = \frac{1}{2}c_1 e^{2t} + 2c_2 e^{-t}$. The I.C. then give
    $c_1 + c_2 = 3$ and $\frac{1}{2}c_1 + 2c_2 = \frac{1}{2}$, which yield
    $c_1 = \frac{11}{3}$, $c_2 = -\frac{2}{3}$. Thus $x_1 = \frac{11}{3}e^{2t} - \frac{2}{3}e^{-t}$ and
    $x_2 = \frac{11}{6}e^{2t} - \frac{4}{3}e^{-t}$.
    Note that for large t, the second term in each solution
    vanishes and we have $x_1 \cong \frac{11}{3}e^{2t}$ and $x_2 \cong \frac{11}{6}e^{2t}$, so that
    $x_1 \cong 2x_2$. This says that the graph will be asymptotic to
    the line $x_1 = 2x_2$ for large t.

9. Eliminating x_2 from the two D.E. yields $x_2 = \frac{4}{3}x_1' - \frac{5}{3}x_1$
    and $x_1'' - 2.5x_1' + x_1 = 0$.

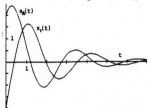

12. Solving the first D.E. for
    x_2 gives $x_2 = \frac{1}{2}x_1' + \frac{1}{4}x_1$ and
    substitution into the second
    D.E. gives $x_1'' + x_1' + \frac{17}{4}x_1 = 0$.
    Thus $x_1 = e^{-t/2}(c_1\cos 2t + c_2\sin 2t)$
    and $x_2 = e^{-t/2}(c_2\cos 2t - c_1\sin 2t)$.
    The I.C. yield $c_1 = -2$ and $c_2 = 2$.

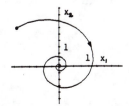

14. If $a_{12} \neq 0$, then solve the first equation for x_2,
     obtaining $x_2 = [x_1' - a_{11}x_1 - g_1(t)]/a_{12}$. Upon substituting
     this expression into the second equation, we have a
     second order linear O.D.E. for x_1. One I.C. is
     $x_1(0) = x_1^0$. The second I.C. is
     $x_2(0) = [x_1'(0) - a_{11}x_1(0) - g_1(0)]/a_{12} = x_2^0$. Solving for
     $x_1'(0)$ gives $x_1'(0) = a_{12}x_2^0 + a_{11}x_1^0 + g_1(0)$. These results
     hold when $a_{11}, \ldots, a_{22}$ are functions of t as long as the
     derivatives exist and $a_{12}(t)$ and $a_{21}(t)$ are not both zero
     on the interval. The initial conditions will involve
     $a_{11}(0)$ and $a_{12}(0)$.

19. Let us number the nodes 1,2, and 3 clockwise beginning
     with the top right node in Figure 7.1.4. Also let I_1,
     I_2, I_3, and I_4 denote the currents through the resistor
     $R = 1$, the inductor $L = 1$, the capacitor $C = \dfrac{1}{2}$, and the
     resistor $R = 2$, respectively. Let V_1, V_2, V_3, and V_4 be
     the corresponding voltage drops. Kirchhoff's first law
     applied to nodes 1 and 2, respectively, gives
     (i) $I_1 - I_2 = 0$ and (ii) $I_2 - I_3 - I_4 = 0$. Kirchhoff's
     second law applied to each loop gives
     (iii) $V_1 + V_2 + V_3 = 0$ and (iv) $V_3 - V_4 = 0$. The current-
     voltage relation through each circuit element yields four
     more equations: (v) $V_1 = I_1$, (vi) $I_2' = V_2$,
     (vii) $(1/2)V_3' = I_3$ and (viii) $V_4 = 2I_4$. We thus have a
     system of eight equations in eight unknowns, and we wish
     to eliminate all of the variables except I_2 and V_3 from
     this system of equations. For example, we can use
     Eqs.(i) and (iv) to eliminate I_1 and V_4 in Eqs.(v) and
     (viii). Then use the new Eqs.(v) and (viii) to eliminate
     V_1 and I_4 in Eqs.(ii) and (iii). Finally, use the new
     Eqs. (ii) and (iii) in Eqs.(vi) and (vii) to obtain
     $I_2' = -I_2 - V_3$, $V_3' = 2I_2 - V_3$. These equations are
     identical (when subscripts on the remaining variables are
     dropped) to the equations given in the text.

Section 7.2, Page 332

1c. Using Eq.(9) and following Example 1 we have

$$AB = \begin{pmatrix} 4 + 2 + 0 & -2 + 10 + 0 & 3 + 0 + 0 \\ 12 - 2 - 6 & -6 + 10 - 1 & 9 + 0 - 2 \\ -8 - 1 + 18 & 4 + 5 + 3 & -6 + 0 + 6 \end{pmatrix},$$

which yields the correct answer.

In problems 10 through 19 the method of row reduction
illustrated in Example 2 can be used to find the inverse
matrix or else to show that none exists. We start with the
original matrix augmented by the indentity matrix, describe a
suitable sequence of elementary row operations, and show the
result of applying these operations.

10. Start with the given matrix augmented by the identity
matrix.
$$\begin{pmatrix} 1 & 4 & . & 1 & 0 \\ & & . & & \\ -1 & 3 & . & 0 & 1 \end{pmatrix}$$

Add 2 times the first row to the second row.
$$\begin{pmatrix} 1 & 4 & . & 1 & 0 \\ & & . & & \\ 0 & 11 & . & 2 & 1 \end{pmatrix}$$
Multiply the second row by (1/11).
$$\begin{pmatrix} 1 & 4 & . & 1 & 0 \\ & & . & & \\ 0 & 1 & . & 2/11 & 1/11 \end{pmatrix}$$
Add (-4) times the second row to the first row.
$$\begin{pmatrix} 1 & 0 & . & 3/11 & -4/11 \\ & & . & & \\ 0 & 1 & . & 2/11 & 1/11 \end{pmatrix}$$

The 2 x 2 matric appearing on the right side of this
augmented matrix is the desired inverse matrix. The
answer can be checked by multiplying it by the given
matrix; the result should be the identity matrix.

12. The augmented matrix in this case is:

$$\begin{pmatrix} 1 & 2 & 3 & . & 1 & 0 & 0 \\ & & & . & & & \\ 2 & 4 & 5 & . & 0 & 1 & 0 \\ & & & . & & & \\ 3 & 5 & 6 & . & 0 & 0 & 1 \end{pmatrix}$$

Add (-2) times the first row to the second row and (-3) times the first row to the third row.

$$\begin{pmatrix} 1 & 2 & 3 & . & 1 & 0 & 0 \\ & & & . & & & \\ 0 & 0 & -1 & . & -2 & 1 & 0 \\ & & & . & & & \\ 0 & -1 & -3 & . & -3 & 0 & 1 \end{pmatrix}$$

Multiply the second and third rows by (-1) and interchange them.

$$\begin{pmatrix} 1 & 2 & 3 & . & 1 & 0 & 0 \\ & & & . & & & \\ 0 & 1 & 3 & . & 3 & 0 & -1 \\ & & & . & & & \\ 0 & 0 & 1 & . & 2 & -1 & 0 \end{pmatrix}$$

Add (-3) times the third row to the first and second

rows.
$$\begin{pmatrix} 1 & 2 & 0 & . & -5 & 3 & 0 \\ & & & . & & & \\ 0 & 1 & 0 & . & -3 & 3 & -1 \\ & & & . & & & \\ 0 & 0 & 1 & . & 2 & -1 & 0 \end{pmatrix}$$

Add (-2) times the second row to the first row.

$$\begin{pmatrix} 1 & 0 & 0 & . & 1 & -3 & 2 \\ & & & . & & & \\ 0 & 1 & 0 & . & -3 & 3 & -1 \\ & & & . & & & \\ 0 & 0 & 1 & . & 2 & -1 & 0 \end{pmatrix}$$

The desired answer appears on the right side of this augmented matrix.

14. Again, start with the given matrix augmented by the

identity matrix.
$$\begin{pmatrix} 1 & 2 & 1 & . & 1 & 0 & 0 \\ \\ -2 & 1 & 8 & . & 0 & 1 & 0 \\ \\ 1 & -2 & -7 & . & 0 & 0 & 1 \end{pmatrix}$$

Add (2) times the first row to the second row and add(-1) times the first row to the third row.
$$\begin{pmatrix} 1 & 2 & 1 & . & 1 & 0 & 0 \\ \\ 0 & 5 & 10 & . & 2 & 1 & 0 \\ \\ 0 & -4 & -8 & . & -1 & 0 & 1 \end{pmatrix}$$

Multiply the second row by (1/5).
$$\begin{pmatrix} 1 & 2 & 1 & . & 1 & 0 & 0 \\ \\ 0 & 1 & 2 & . & 2/5 & 1/5 & 0 \\ \\ 0 & -4 & -8 & . & -1 & 0 & 1 \end{pmatrix}$$

Add (-2) times the second row to the first row and add (4) times the second row to the third row.
$$\begin{pmatrix} 1 & 0 & -3 & . & 1/5 & -2/5 & 0 \\ \\ 0 & 1 & 2 & . & 2/5 & 1/5 & 0 \\ \\ 0 & 0 & 0 & . & 3/5 & 4/5 & 1 \end{pmatrix}$$

Since the element in the third row and third column is zero, no further reduction can be performed. Since it cannot be reduced to the identity matrix, the given matrix is singular.

23. $\underset{\sim}{x}' = \begin{pmatrix} 4 \\ 2 \end{pmatrix} 2e^{2t} = \begin{pmatrix} 8 \\ 4 \end{pmatrix} e^{2t};$

$\begin{pmatrix} 3 & -2 \\ 2 & -2 \end{pmatrix} \underset{\sim}{x} = \begin{pmatrix} 3 & -2 \\ 2 & -2 \end{pmatrix} \begin{pmatrix} 4 \\ 2 \end{pmatrix} e^{2t} = \begin{pmatrix} 12-4 \\ 8-4 \end{pmatrix} e^{2t} = \begin{pmatrix} 8 \\ 4 \end{pmatrix} e^{2t}.$

Section 7.3, Page 343

1. Form the augmented matrix, as in Example 1, and use row
    reduction.
$$\begin{pmatrix} 1 & 0 & -1 & . & 0 \\ & & & . & \\ 3 & 1 & 1 & . & 1 \\ & & & . & \\ -1 & 1 & 2 & . & 2 \end{pmatrix}$$
    Add (-3) times the first row to the second and add the
    first row to the third.
$$\begin{pmatrix} 1 & 0 & -1 & . & 0 \\ & & & . & \\ 0 & 1 & 4 & . & 1 \\ & & & . & \\ 0 & 1 & 1 & . & 2 \end{pmatrix}$$
    Add (-1) times the second row to the third.
$$\begin{pmatrix} 1 & 0 & -1 & . & 0 \\ & & & . & \\ 0 & 1 & 4 & . & 1 \\ & & & . & \\ 0 & 0 & -3 & . & 1 \end{pmatrix}$$
    The third row is equivalent to $- 3x_3 = 1$ or $x_3 = - 1/3$.
    Likewise the second row is equivalent to $x_2 + 4x_3 = 1$, so
    $x_2 = 7/3$. Finally, from the first row, $x_1 - x_3 = 0$, so
    $x_1 = - 1/3$. The answer $x_1 = - 1/3$, $x_2 = 7/3$ and
    $x_3 = - 1/3$, of course, could be checked by substituting
    into the original equations.

2. Form the augmented matrix and use row reduction to obtain
$$\begin{pmatrix} 1 & 2 & -1 & . & 1 \\ & & & . & \\ 0 & -3 & 3 & . & -1 \\ & & & . & \\ 0 & 0 & 0 & . & 1 \end{pmatrix}.$$
    The last row corresponds to the equation
    $0x_1 + 0x_2 + 0x_3 = 1$, and there is no choice of x_1, x_2, and
    x_3 that satisfies this equation. Hence the given system
    of equations has no solution.

3.   Form the augmented matrix and use row reduction.

$$\begin{pmatrix} 1 & 2 & -1 & . & 2 \\ & & & . & \\ 2 & 1 & 1 & . & 1 \\ & & & . & \\ 1 & -1 & 2 & . & -1 \end{pmatrix}$$

Add (-2) times the first row to the second and add (-1) times the first row to the third.

$$\begin{pmatrix} 1 & 2 & -1 & . & 2 \\ & & & . & \\ 0 & -3 & 3 & . & -3 \\ & & & . & \\ 0 & -3 & 3 & . & -3 \end{pmatrix}$$

Add (-1) times the second row to the third row and then multiply the second row by $(-1/3)$.

$$\begin{pmatrix} 1 & 2 & -1 & . & 2 \\ & & & . & \\ 0 & 1 & -1 & . & 1 \\ & & & . & \\ 0 & 0 & 0 & . & 0 \end{pmatrix}$$

Since the last row has only zero entries, it may be dropped. The second row corresponds to the equation $x_2 - x_3 = 1$. We can assign an arbitrary value to either x_2 or x_3 and use this equation to solve for the other. For example, let $x_3 = c$, where c is arbitrary. Then $x_2 = 1 + c$. The first row corresponds to the equation $x_1 + 2x_2 - x_3 = 2$, so $x_1 = 2 - 2x_2 + x_3 = 2 - 2(1+c)+c = -c$.

6.   To determine whether the given set of vectors is linearly independent we must solve the system
$c_1 \underset{\sim}{x}^{(1)} + c_2 \underset{\sim}{x}^{(2)} + c_3 \underset{\sim}{x}^{(3)} = \underset{\sim}{0}$ for c_1, c_2, and c_3.

Form the augmented matrix and use row reduction.

$$\begin{pmatrix} 1 & 0 & 1 & . & 0 \\ & & & . & \\ 1 & 1 & 0 & . & 0 \\ & & & . & \\ 0 & 1 & 1 & . & 0 \end{pmatrix}$$

Add (-1) times the first row to the second.

$$\begin{pmatrix} 1 & 0 & 1 & . & 0 \\ & & & . & \\ 0 & 1 & -1 & . & 0 \\ & & & . & \\ 0 & 1 & 1 & . & 0 \end{pmatrix}$$

Add (-1) times the second row to the third.

$$\begin{pmatrix} 1 & 0 & 1 & . & 0 \\ & & & . & \\ 0 & 1 & -1 & . & 0 \\ & & & . & \\ 0 & 0 & 2 & . & 0 \end{pmatrix}$$

From the third row we have $c_3 = 0$. Then from the second row, $c_2 - c_3 = 0$, so $c_2 = 0$. Finally from the first row $c_1 + c_3 = 0$, so $c_1 = 0$. Since $c_1 = c_2 = c_3 = 0$, we conclude that the given vectors are linearly independent.

8. As in Problem 6 we wish to solve the system
$c_1 \underset{\sim}{x}^{(1)} + c_2 \underset{\sim}{x}^{(2)} + c_3 \underset{\sim}{x}^{(3)} + c_4 \underset{\sim}{x}^{(4)} = \underset{\sim}{0}$ for c_1, c_2, c_3, and c_4.

Form the augmented matrix and use row reduction.

$$\begin{pmatrix} 1 & -1 & -2 & -3 & . & 0 \\ & & & & . & \\ 2 & 0 & -1 & 0 & . & 0 \\ & & & & . & \\ 2 & 3 & 1 & -1 & . & 0 \\ & & & & . & \\ 3 & 1 & 0 & 3 & . & 0 \end{pmatrix}$$

Add (-2) times the first row to the second, add (-2) times the first row to the third, and add (-3) times the first row to the fourth.

$$\begin{pmatrix} 1 & -1 & -2 & -3 & . & 0 \\ & & & & . & \\ 0 & 2 & 3 & 6 & . & 0 \\ & & & & . & \\ 0 & 5 & 5 & 5 & . & 0 \\ & & & & . & \\ 0 & 4 & 6 & 12 & . & 0 \end{pmatrix}$$

Multiply the second row by (1/2) and then add (-5) times the second row to the third and add (-4) times the second row to the fourth.

$$\begin{pmatrix} 1 & -1 & -2 & -3 & . & 0 \\ & & & & . & \\ 0 & 1 & 3/2 & 3 & . & 0 \\ & & & & . & \\ 0 & 0 & -5/2 & -10 & . & 0 \\ & & & & . & \\ 0 & 0 & 0 & 0 & . & 0 \end{pmatrix}$$

The third row is equivalent to the equation $c_3 + 4c_4 = 0$. One way to satisfy this equation is by choosing $c_4 = -1$; then $c_3 = 4$. From the second row we have $c_2 = - (3/2)c_3 - 3c_4 = - 6 + 3 = -3$. Then, from the first row, $c_1 = c_2 + 2c_3 + 3c_4 = -3 + 8 - 3 = 2$. Hence the given vectors are linearly dependent, and satisfy $2\underset{\sim}{x}^{(1)} - 3\underset{\sim}{x}^{(2)} + 4\underset{\sim}{x}^{(3)} - \underset{\sim}{x}^{(4)} = 0$.

14. Let $t = t_0$ be a fixed value of t in the interval $0 \leq t \leq 1$. To determine whether $\underset{\sim}{x}^{(1)}(t_0)$ and $\underset{\sim}{x}^{(2)}(t_0)$ are linearly dependent we must solve $c_1\underset{\sim}{x}^{(1)}(t_0) + c_2\underset{\sim}{x}^{(2)}(t_0) = 0$. We have the augmented matrix

$$\begin{pmatrix} e^{t_0} & 1 & . & 0 \\ & & . & \\ & & . & \\ t_0 e^{t_0} & t_0 & . & 0 \end{pmatrix}$$

and by row reduction we obtain $\begin{pmatrix} e^{t_0} & 1 & . & 0 \\ & & . & \\ & & . & \\ 0 & 0 & . & 0 \end{pmatrix}$.

Thus, for example, we can choose $c_1 = 1$ and $c_2 = -e^{t_0}$, and hence the given vectors are linearly dependent at t_0. Since t_0 is arbitrary the vectors are linearly dependent at each point in the interval. However, there is no linear relation between $\underset{\sim}{x}^{(1)}$ and $\underset{\sim}{x}^{(2)}$ that is valid throughout the interval $0 \leq t \leq 1$. For example, if $t_1 \neq t_0$, and if c_1 and c_2 are chosen as above, then $c_1 \underset{\sim}{x}^{(1)}(t_1) + c_2 \underset{\sim}{x}^{(2)}(t_1)$

$$= \begin{pmatrix} e^{t_1} \\ t_1 e^{t_1} \end{pmatrix} + -e^{t_0} \begin{pmatrix} 1 \\ t_1 \end{pmatrix} = \begin{pmatrix} e^{t_1} - e^{t_0} \\ t_1 e^{t_1} - t_1 e^{t_0} \end{pmatrix} \neq \begin{pmatrix} 0 \\ 0 \end{pmatrix}.$$

Hence the given vectors must be linearly independent on $0 \leq t \leq 1$. In fact, the same argument applies to any interval.

15. To find the eigenvalues and eigenvectors of the given matrix we must solve $\begin{pmatrix} 5-\lambda & -1 \\ 3 & 1-\lambda \end{pmatrix} \begin{pmatrix} x_1 \\ x_2 \end{pmatrix} = \begin{pmatrix} 0 \\ 0 \end{pmatrix}$. The determinant of coefficients is $(5-\lambda)(1-\lambda) - (-1)(3) = 0$, or $\lambda^2 - 6\lambda + 8 = 0$. Hence $\lambda_1 = 2$ and $\lambda_2 = 4$ are the eigenvalues. The eigenvector corresponding to λ_1 must satisfy $\begin{pmatrix} 3 & -1 \\ 3 & -1 \end{pmatrix} \begin{pmatrix} x_1 \\ x_2 \end{pmatrix} = \begin{pmatrix} 0 \\ 0 \end{pmatrix}$, or $3x_1 - x_2 = 0$. If we let $x_1 = 1$, then $x_2 = 3$ and the eigenvector is $\underset{\sim}{x}^{(1)} = \begin{pmatrix} 1 \\ 3 \end{pmatrix}$, or any constant multiple of this vector. Similarly, the eigenvector corresponding to λ_2 must satisfy $\begin{pmatrix} 1 & -1 \\ 3 & -3 \end{pmatrix} \begin{pmatrix} x_1 \\ x_2 \end{pmatrix} = \begin{pmatrix} 0 \\ 0 \end{pmatrix}$, or $x_1 - x_2 = 0$. Hence $\underset{\sim}{x}^{(2)} = \begin{pmatrix} 1 \\ 1 \end{pmatrix}$, or a multiple thereof.

18. The given matrix is Hermitian so we know in advance that its eigenvalues are real. To find the eigenvalues and

eigenvectors we must solve $\begin{pmatrix} 1-\lambda & i \\ -i & 1-\lambda \end{pmatrix} \begin{pmatrix} x_1 \\ x_2 \end{pmatrix} = \begin{pmatrix} 0 \\ 0 \end{pmatrix}$. The

determinant of coefficients is $(1-\lambda)^2 - i(-i) = \lambda^2 - 2\lambda$, so the eigenvalues are $\lambda_1 = 0$ and $\lambda_2 = 2$; observe that they are indeed real even though the given matrix has imaginary entries. The eigenvector corresponding to λ_1

must satisfy $\begin{pmatrix} 1 & i \\ -i & 1 \end{pmatrix} \begin{pmatrix} x_1 \\ x_2 \end{pmatrix} = \begin{pmatrix} 0 \\ 0 \end{pmatrix}$, or $x_1 + ix_2 = 0$. Note

that the second equation $-ix_1 + x_2 = 0$ is a multiple of the first. If $x_1 = 1$, then $x_2 = i$, and the eigenvector

is $\underset{\sim}{x}^{(1)} = \begin{pmatrix} 1 \\ i \end{pmatrix}$. In a similar way we find that the

eigenvector associated with λ_2 is $\underset{\sim}{x}^{(2)} = \begin{pmatrix} 1 \\ -1 \end{pmatrix}$.

20. The eigenvalues and eigenvectors satisfy
$\begin{pmatrix} 1-\lambda & -4 \\ 4 & -7-\lambda \end{pmatrix} \begin{pmatrix} x_1 \\ x_2 \end{pmatrix} = \begin{pmatrix} 0 \\ 0 \end{pmatrix}$. The determinant of coefficients
is $(1-\lambda)(-7-\lambda) - (-4)4 = \lambda^2 + 6\lambda + 9$, so $\lambda_1 = \lambda_2 = -3$.
Thus -3 is a double eigenvalue. The corresponding
eigenvectors satisfy $\begin{pmatrix} 4 & -4 \\ 4 & -4 \end{pmatrix} \begin{pmatrix} x_1 \\ x_2 \end{pmatrix} = \begin{pmatrix} 0 \\ 0 \end{pmatrix}$. Hence

$x_1 - x_2 = 0$, so $x_1 = x_2$ and $\underset{\sim}{x}^{(1)} = \begin{pmatrix} 1 \\ 1 \end{pmatrix}$, or any multiple

thereof. There are no other linearly independent eigenvectors in this problem since both equations in this last set are identical.

24. Since the given matrix is real and symmetric, we know that the eigenvalues are real. Further, even if there are repeated eigenvalues, there will be a full set of three linearly independent eigenvectors. To find the eigenvalues and eigenvectors we must solve
$\begin{pmatrix} 3-\lambda & 2 & 4 \\ 2 & -\lambda & 2 \\ 4 & 2 & 3-\lambda \end{pmatrix} \begin{pmatrix} x_1 \\ x_2 \\ x_3 \end{pmatrix} = \begin{pmatrix} 0 \\ 0 \\ 0 \end{pmatrix}$. The determinant of

coefficients is $(3-\lambda)[-\lambda(3-\lambda)-4] - 2[2(3-\lambda) -8] + 4[4+4\lambda]$
$= -\lambda^3 + 6\lambda^2 + 15\lambda + 8$. Setting this equal to zero and
solving we find $\lambda_1 = \lambda_2 = -1$, $\lambda_3 = 8$. The eigenvectors
corresponding to λ_1 and λ_2 must satisfy

$$\begin{pmatrix} 4 & 2 & 4 \\ 2 & 1 & 2 \\ 4 & 2 & 4 \end{pmatrix} \begin{pmatrix} x_1 \\ x_2 \\ x_3 \end{pmatrix} = \begin{pmatrix} 0 \\ 0 \\ 0 \end{pmatrix};$$ hence there is only the single

relation $2x_1 + x_2 + 2x_3 = 0$ to be satisfied.
Consequently, two of the variables can be selected
arbitrarily and the third is then determined by this
equation. For example, if $x_1 = 1$ and $x_3 = 1$, then

$x_2 = -4$, and we obtain the eigenvector $x^{(1)} = \begin{pmatrix} 1 \\ -4 \\ 1 \end{pmatrix}$.

Similarly, if $x_1 = 1$ and $x_2 = 0$, then $x_3 = -1$, and we

have the eigenvector $x^{(2)} = \begin{pmatrix} 1 \\ 0 \\ -1 \end{pmatrix}$, which is linearly

independent of $x^{(1)}$. There are many other choices that

could have been made; however, by Eq.(28) there can be no
more than two linearly independent eigenvectors
corresponding to the eigenvalue -1. To find the
eigenvector corresponding to λ_3 we must solve

$$\begin{pmatrix} -5 & 2 & 4 \\ 2 & -8 & 2 \\ 4 & 2 & -5 \end{pmatrix} \begin{pmatrix} x_1 \\ x_2 \\ x_3 \end{pmatrix} = \begin{pmatrix} 0 \\ 0 \\ 0 \end{pmatrix}.$$ By row reduction we can obtain

the equivalent system $x_1 - 4x_2 + x_3 = 0$, $2x_2 - x_3 = 0$.
Since there are two equations to satisfy only one
variable can be assigned an arbitrary value. If we let
$x_2 = 1$, then $x_3 = 2$ and $x_1 = 2$, so we find that

$$x^{(3)} = \begin{pmatrix} 2 \\ 1 \\ 2 \end{pmatrix}.$$

29a. From the discussion leading to Eq.(44) we find T is

composed of the eigenvectors found in Problem 15 and thus

$$T = \begin{pmatrix} 1 & 1 \\ 3 & 1 \end{pmatrix}.$$ To verify Eq.(46) we find

$$T^{-1} = \begin{pmatrix} -1/2 & 1/2 \\ 3/2 & -1/2 \end{pmatrix}$$ and calculate $T^{-1}AT$ to find that

$$T^{-1}AT = \begin{pmatrix} 2 & 0 \\ 0 & 4 \end{pmatrix},$$ as indicated in Eq.(45).

Section 7.4, Page 349

2a. From Eq.(10) we have

$$W = \begin{vmatrix} x_1^{(1)} & x_1^{(2)} \\ x_2^{(1)} & x_2^{(1)} \end{vmatrix} = x_1^{(1)} x_2^{(2)} - x_2^{(1)} x_1^{(2)}.$$ Taking the

derivative of these two products yields four terms which may be written as

$$\frac{dw}{dt} = [\frac{dx_1^{(1)}}{dt} x_2^{(2)} - x_2^{(1)} \frac{dx_1^{(2)}}{dt}] + [x_1^{(1)} \frac{dx_2^{(2)}}{dt} - \frac{dx_2^{(1)}}{dt} x_1^{(2)}].$$

The terms in the square brackets can now be recognized as the respective determinants appearing in the desired solution. A similar result was mentioned in Problem 22 of Section 4.1.

2b. If $x^{(1)}$ is substituted into Eq.(3) we have

$$\frac{dx_1^{(1)}}{dt} = p_{11} x_1^{(1)} + p_{12} x_2^{(1)}$$

$$\frac{dx_2^{(1)}}{dt} = p_{21} x_1^{(1)} + p_{22} x_2^{(1)}.$$

Substituting the first equation above and its counterpart for $x^{(2)}$ into the first determinant appearing in dW/dt

and evaluating the result yields $p_{11} \begin{vmatrix} x_1^{(1)} & x_1^{(2)} \\ x_2^{(1)} & x_2^{(2)} \end{vmatrix} = p_{11}W.$

Similarly, the second determinant in dW/dt is evaluated as $p_{22}W$, yielding the desired result.

6a. From Eq.(10) $W = \begin{vmatrix} t & t^2 \\ 1 & 2t \end{vmatrix}.$

6d. To obtain the system satisfied by $x^{(1)}$ and $x^{(2)}$ we proceed in a fashion similar to Problems 31 to 36 of Section 3.1. The general solution of the system can be written as

$$\underset{\sim}{x} = c_1\underset{\sim}{x}^{(1)} + c_2\underset{\sim}{x}^{(2)}, \text{ or } \begin{pmatrix} x_1 \\ x_2 \end{pmatrix} = c_1 \begin{pmatrix} t \\ 1 \end{pmatrix} + c_2 \begin{pmatrix} t^2 \\ 2t \end{pmatrix}.$$

Taking the derivative we obtain $\begin{pmatrix} x_1' \\ x_2' \end{pmatrix} = c_1 \begin{pmatrix} 1 \\ 0 \end{pmatrix} + c_2 \begin{pmatrix} 2t \\ 2 \end{pmatrix}.$

Solving this last system for c_1 and c_2 we find
$c_1 = x_1' - tx_2'$ and $c_2 = x_2'/2$. Thus

$$\begin{pmatrix} x_1 \\ x_2 \end{pmatrix} = (x_1' - tx_2') \begin{pmatrix} t \\ 1 \end{pmatrix} + \frac{x_2'}{2} \begin{pmatrix} t^2 \\ 2t \end{pmatrix}, \text{ which yields}$$

$x_1 = tx_1' - \dfrac{t^2}{2} x_2'$ and $x_2 = x_1'$. Writing this system in

matrix form we have $\underset{\sim}{x} = \begin{pmatrix} t - t^2/2 \\ 1 \quad\quad 0 \end{pmatrix} \underset{\sim}{x}'.$ Finding the

inverse of the matrix multiplying $\underset{\sim}{x}'$ yields the desired
solution.

Section 7.5, Page 356

1. Assuming that there are solutions of the form $\underset{\sim}{x} = \underset{\sim}{\xi}e^{rt}$,
    we substitute into the D.E. to find
    $r\underset{\sim}{\xi}e^{rt} = \begin{pmatrix} 3 & -2 \\ 2 & -2 \end{pmatrix} \underset{\sim}{\xi}e^{rt}.$ Since $\underset{\sim}{\xi} = I\underset{\sim}{\xi} = \begin{pmatrix} 1 & 0 \\ 0 & 1 \end{pmatrix} \underset{\sim}{\xi}$, we can

    write this equation as $\begin{pmatrix} 3 & -2 \\ 2 & -2 \end{pmatrix} \underset{\sim}{\xi} - r \begin{pmatrix} 1 & 0 \\ 0 & 1 \end{pmatrix} \underset{\sim}{\xi} = \underset{\sim}{0}$ and

    thus we must solve $\begin{pmatrix} 3-r & -2 \\ 2 & -2-r \end{pmatrix} \begin{pmatrix} \xi_1 \\ \xi_2 \end{pmatrix} = \begin{pmatrix} 0 \\ 0 \end{pmatrix}$ for r, ξ_1, ξ_2.

    The determinant of the coefficients is
    $(3-r)(-2-r) + 4 = r^2 - r - 2$, so the eigenvalues are
    $r = -1, 2$. The eigenvector corresponding to $r = -1$
    satisfies
    $\begin{pmatrix} 4 & -2 \\ 2 & -1 \end{pmatrix} \begin{pmatrix} \xi_1 \\ \xi_2 \end{pmatrix} = \begin{pmatrix} 0 \\ 0 \end{pmatrix}$, which yields $2\xi_1 - \xi_2 = 0$. Thus

    $\underset{\sim}{x}^{(1)}(t) = \underset{\sim}{\xi}^{(1)}e^{-t} = \begin{pmatrix} 1 \\ 2 \end{pmatrix} e^{-t}$, where we have set $\xi_1 = 1$.

    (Any other non zero choice would also work). In a

    similar fashion, for $r = 2$, we have $\begin{pmatrix} 1 & -2 \\ 2 & -4 \end{pmatrix} \begin{pmatrix} \xi_1 \\ \xi_2 \end{pmatrix} = \begin{pmatrix} 0 \\ 0 \end{pmatrix}$, or $\xi_1 - 2\xi_2 = 0$. Hence $\underset{\sim}{x}^{(2)}(t) = \underset{\sim}{\xi}^{(2)} e^{2t} = \begin{pmatrix} 2 \\ 1 \end{pmatrix} e^{2t}$ by

setting $\xi_2 = 1$. The general solution is then
$\underset{\sim}{x} = c_1 \underset{\sim}{x}^{(1)}(t) + c_2 \underset{\sim}{x}^{(2)}(t)$. To sketch the trajectories we
follow the steps illustrated in Examples 1 and 2.

Setting $c_2 = 0$ we have $\underset{\sim}{x} = \begin{pmatrix} x_1 \\ x_2 \end{pmatrix} = c_1 \begin{pmatrix} 1 \\ 2 \end{pmatrix} e^{-t}$ or $x_1 = c_1 e^{-t}$

and $x_2 = 2c_1 e^{-t}$ and thus one asymptote is given by

$x_2 = 2x_1$. In a similar
fashion $c_1 = 0$ gives
$x_2 = (1/2)x_1$ as asecond
asymptote. Since the
roots differ in sign,
the trajectories for
this problem are similar
in nature to those in
Example 1. For $c_2 \neq 0$,
all solutions will be

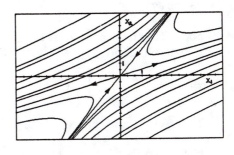

asymptotic to $x_2 = (1/2)x_1$ as $t \to \infty$. For $c_2 = 0$, the
solution approaches the origin along the line $x_2 = 2x_1$.

5. Proceeding as in Problem 1 we assume a solution of the
    form $\underset{\sim}{x} = \underset{\sim}{\xi} e^{rt}$, where r, ξ_1, ξ_2 must now satisfy

    $\begin{pmatrix} -2-r & 1 \\ 1 & -2-r \end{pmatrix} \begin{pmatrix} \xi_1 \\ \xi_2 \end{pmatrix} = \begin{pmatrix} 0 \\ 0 \end{pmatrix}$. Evaluating the determinant of the

    coefficients set equal to zero yields $r = -1, -3$ as the
    eigenvalues. For $r = -1$ we find $\xi_1 = \xi_2$ and thus

    $\underset{\sim}{\xi}^{(1)} = \begin{pmatrix} 1 \\ 1 \end{pmatrix}$ and for $r = -3$ we find $\xi_2 = -\xi_1$ and hence

    $\underset{\sim}{\xi}^{(2)} = \begin{pmatrix} 1 \\ -1 \end{pmatrix}$. The general solution is then

    $\underset{\sim}{x} = c_1 \begin{pmatrix} 1 \\ 1 \end{pmatrix} e^{-t} + c_2 \begin{pmatrix} 1 \\ -1 \end{pmatrix} e^{-3t}$. Since there are two negative

    eigenvalues, we would expect the trajectories to be
    similar to those of Example 2.
    Setting $c_2 = 0$ and eliminating
    t (as in Problem 1) we find

    that $\begin{pmatrix} 1 \\ 1 \end{pmatrix} e^{-t}$ approaches the

    origin along the line $x_2 = x_1$.

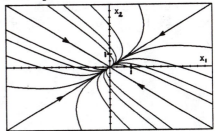

Similarly $\begin{pmatrix} 1 \\ -1 \end{pmatrix} e^{-3t}$ approaches the origin along the line $x_2 = -x_1$. As long as $c_1 \neq 0$, all trajectories approach the origin asymptotic to $x_2 = x_1$ for $c_1 = 0$, the trajectory approaches the origin along $x_2 = -x_1$, as shown in the graph.

7. Again assuming $\underset{\sim}{x} = \underset{\sim}{\xi} e^{rt}$ we find that r, ξ_1, ξ_2 must satisfy $\begin{pmatrix} 4-r & -3 \\ 8 & -6-r \end{pmatrix} \begin{pmatrix} \xi_1 \\ \xi_2 \end{pmatrix} = \begin{pmatrix} 0 \\ 0 \end{pmatrix}$. The determinant of the coefficients set equal to zero yields $r = 0, -2$. For $r = 0$ we find $4\xi_1 = 3\xi_2$. Choosing $\xi_2 = 4$ we find $\xi_1 = 3$ and thus $\underset{\sim}{\xi}^{(1)} = \begin{pmatrix} 3 \\ 4 \end{pmatrix}$. Similarly for $r = -2$ we have $\underset{\sim}{\xi}^{(2)} = \begin{pmatrix} 1 \\ 2 \end{pmatrix}$ and thus $\underset{\sim}{x} = c_1 \begin{pmatrix} 3 \\ 4 \end{pmatrix} + c_2 \begin{pmatrix} 1 \\ 2 \end{pmatrix} e^{-2t}$. To sketch the trajectories, note that the general solution is equivalent to the simultaneous equations $x_1 = 3c_1 + c_2 e^{-2t}$ and $x_2 = 4c_1 + 2c_2 e^{-2t}$. Solving the first equation for $c_2 e^{-2t}$ and substituting into the second yields $x_2 = 2x_1 - 2c_1$ and thus the trajectories are parallel straight lines.

15. The eigenvalues and eigenvectors of the coefficient matrix satisfy $\begin{pmatrix} 1-r & -1 & 4 \\ 3 & 2-r & -1 \\ 2 & 1 & -1-r \end{pmatrix} \begin{pmatrix} \xi_1 \\ \xi_2 \\ \xi_3 \end{pmatrix} = \begin{pmatrix} 0 \\ 0 \\ 0 \end{pmatrix}$. The determinant of coefficients set equal to zero reduces to $r^3 - 2r^2 - 5r + 6 = 0$, so the eigenvalues are $r_1 = 1$, $r_2 = -2$, and $r_3 = 3$. The eigenvector corresponding to r_1 must satisfy $\begin{pmatrix} 0 & -1 & 4 \\ 3 & 1 & -1 \\ 2 & 1 & -2 \end{pmatrix} \begin{pmatrix} \xi_1 \\ \xi_2 \\ \xi_3 \end{pmatrix} = \begin{pmatrix} 0 \\ 0 \\ 0 \end{pmatrix}$.
Using row reduction we obtain the equivalent system $\xi_1 + \xi_3 = 0$, $\xi_2 - 4\xi_3 = 0$. Letting $\xi_1 = 1$, it follows that $\xi_3 = -1$ and $\xi_2 = -4$, so $\underset{\sim}{\xi}^{(1)} = \begin{pmatrix} 1 \\ -4 \\ -1 \end{pmatrix}$. In a similar way the eigenvectors corresponding to r_2 and r_3 are found

to be $\xi^{(2)} = \begin{pmatrix} 1 \\ -1 \\ -1 \end{pmatrix}$ and $\xi^{(3)} = \begin{pmatrix} 1 \\ 2 \\ 1 \end{pmatrix}$, respectively. Thus the

general solution of the given D.E. is

$$x = c_1 \begin{pmatrix} 1 \\ -4 \\ -1 \end{pmatrix} e^t + c_2 \begin{pmatrix} 1 \\ -1 \\ -1 \end{pmatrix} e^{-2t} + c_3 \begin{pmatrix} 1 \\ 2 \\ 1 \end{pmatrix} e^{3t}.$$ Notice that the

"trajectories" of this solution would lie in the x_1 x_2 x_3
three dimentional space.

17. The eigenvalues and eigenvectors of the coefficient

matrix are found to be $r_1 = -1$, $\xi^{(1)} = \begin{pmatrix} 1 \\ 1 \end{pmatrix}$ and $r_2 = 3$,

$\xi^{(2)} = \begin{pmatrix} 1 \\ 5 \end{pmatrix}$. Thus the general solution of the given D.E.

is $x = c_1 \begin{pmatrix} 1 \\ 1 \end{pmatrix} e^{-t} + c_2 \begin{pmatrix} 1 \\ 5 \end{pmatrix} e^{3t}$. The I.C. yields the

system of equations $c_1 \begin{pmatrix} 1 \\ 1 \end{pmatrix} + c_2 \begin{pmatrix} 1 \\ 5 \end{pmatrix} = \begin{pmatrix} 1 \\ 3 \end{pmatrix}$. The augmented

matrix of this system is $\begin{pmatrix} 1 & 1 & . & 1 \\ & & . & \\ 1 & 5 & . & 3 \end{pmatrix}$ and by row reduction

we obtain $\begin{pmatrix} 1 & 1 & . & 1 \\ & & . & \\ 0 & 1 & .1/2 \end{pmatrix}$. Thus $c_2 = 1/2$ and $c_1 = 1/2$.

Substituting these values in the general solution gives
the solution of the I.V.P. As $t \to \infty$, the solution

becomes asymptotic to $x = \frac{1}{2} \begin{pmatrix} 1 \\ 5 \end{pmatrix} e^{3t}$, or $x_2 = 5x_1$.

21. Substituting $x = \xi t^r$ into the D.E. we obtain

$r\xi t^r = \begin{pmatrix} 2 & -1 \\ 3 & -2 \end{pmatrix} \xi t^r$. For $t \neq 0$ this equation can be

written as $\begin{pmatrix} 2-r & -1 \\ 3 & -2-r \end{pmatrix} \begin{pmatrix} \xi_1 \\ \xi_2 \end{pmatrix} = \begin{pmatrix} 0 \\ 0 \end{pmatrix}$. The eigenvalues and

eigenvectors are $r_1 = 1$, $\xi^{(1)} = \begin{pmatrix} 1 \\ 1 \end{pmatrix}$ and $r_2 = -1$,

$\xi^{(2)} = \begin{pmatrix} 1 \\ 3 \end{pmatrix}$. Substituting these in the assumed form we

obtain the general solution $x = c_1 \begin{pmatrix} 1 \\ 1 \end{pmatrix} t + c_2 \begin{pmatrix} 1 \\ 3 \end{pmatrix} t^{-1}$.

28. The eigenvalues and eigenvectors of the coefficient matrix were found in Problem 20 of Section 7.3, namely,

$r_1 = r_2 = -3$, $\xi^{(1)} = \begin{pmatrix} 1 \\ 1 \end{pmatrix}$. Thus one solution of the given

D.E. is $x^{(1)}(t) = \begin{pmatrix} 1 \\ 1 \end{pmatrix} e^{-3t}$ but there is no second solution

of the form $x = \xi e^{rt}$. To use the method of reduction of

order, as shown in Problem 27, let $x = \begin{pmatrix} 1 & e^{-3t} \\ 0 & e^{-3t} \end{pmatrix} y$. Then

it follows that $x' = \begin{pmatrix} 1 & e^{-3t} \\ 0 & e^{-3t} \end{pmatrix} y' + \begin{pmatrix} 0 & -3e^{-3t} \\ 0 & -3e^{-3t} \end{pmatrix} y$ and that

$Ax = \begin{pmatrix} 1 & -3e^{-3t} \\ 4 & -3e^{-3t} \end{pmatrix} y$. Substituting into the D.E., we

obtain $\begin{pmatrix} 1 & e^{-3t} \\ 0 & e^{-3t} \end{pmatrix} y' = \begin{pmatrix} 1 & 0 \\ 4 & 0 \end{pmatrix} y$. Multiplying both sides of

this last equation by $\begin{pmatrix} 1 & -1 \\ 0 & e^{3t} \end{pmatrix}$, the inverse of the matrix

on the left, we obtain $y' = \begin{pmatrix} -3 & 0 \\ 4e^{3t} & 0 \end{pmatrix} y$. This equation has

a variable matrix, so it can't be solved by the methods of this section. However, we can write the equation in scalar form as $y_1' = -3y_1$ and $y_2' = 4e^{3t}y_1$. Solving the first equation we get $y_1 = k_1 e^{-3t}$ and thus $y_2' = 4k_1$, which

yields $y_2 = 4k_1 t + k_2$. Consequently $y = k_1 \begin{pmatrix} e^{-3t} \\ 4t \end{pmatrix} + k_2 \begin{pmatrix} 0 \\ 1 \end{pmatrix}$

and hence $x = \begin{pmatrix} 1 & e^{-3t} \\ 0 & e^{-3t} \end{pmatrix} y = \begin{pmatrix} k_1 + k_2 \\ k_2 \end{pmatrix} e^{-3t} + 4k_1 \begin{pmatrix} 1 \\ 1 \end{pmatrix} te^{-3t}$.

By setting $k_1 = 1/4$ and $k_2 = -1/4$ we obtain the form of the second solution given in the text. See Problem 7 of Section 7.7 for another solution of this problem.

Section 7.6, Page 364

1. We assume a solution of the form $x = \xi e^{rt}$ thus r and ξ

    are solutions of $\begin{pmatrix} 3-r & -2 \\ 4 & -1-r \end{pmatrix} \begin{pmatrix} \xi_1 \\ \xi_2 \end{pmatrix} = \begin{pmatrix} 0 \\ 0 \end{pmatrix}$. The determinant of

    coefficients is $(r^2-2r-3) + 8 = r^2 - 2r + 5$, so the
    eigenvalues are $r = 1 \pm 2i$. The eigenvector

    corresponding to $1 + 2i$ satisfies $\begin{pmatrix} 2-2i & -2 \\ 4 & -2-2i \end{pmatrix} \begin{pmatrix} \xi_1 \\ \xi_2 \end{pmatrix} = \begin{pmatrix} 0 \\ 0 \end{pmatrix}$,

    or $(2-2i)\xi_1 - 2\xi_2 = 0$. If $\xi_1 = 1$, then $\xi_2 = 1-i$ and

    $\xi^{(1)} = \begin{pmatrix} 1 \\ 1-i \end{pmatrix}$ and thus one

    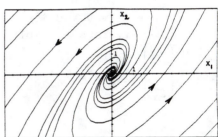

    complex-valued solution
    of the D.E. is

    $x^{(1)}(t) = \begin{pmatrix} 1 \\ 1-i \end{pmatrix} e^{(1+2i)t}$.

    To find real-valued solutions
    (see Eqs. 8 and 9) we take
    the real and imaginary parts,
    respectively of $x^{(1)}(t)$.

    Thus $x^{(1)}(t) = \begin{pmatrix} 1 \\ 1-i \end{pmatrix} e^t(\cos 2t + i\sin 2t)$

    $= e^t \begin{pmatrix} \cos 2t + i\sin 2t \\ \cos 2t + \sin 2t + i(\sin 2t - \cos 2t) \end{pmatrix}$

    $= e^t \begin{pmatrix} \cos 2t \\ \cos 2t + \sin 2t \end{pmatrix} + ie^t \begin{pmatrix} \sin 2t \\ \sin 2t - \cos 2t \end{pmatrix}$

    Hence the general solution of the D.E. is

    $x = c_1 e^t \begin{pmatrix} \cos 2t \\ \cos 2t + \sin 2t \end{pmatrix} + c_2 e^t \begin{pmatrix} \sin 2t \\ \sin 2t - \cos 2t \end{pmatrix}$. The

    solutions spiral to ∞ as $t \to \infty$ due to the e^t terms.

7. The eigenvalues and eigenvectors of the coefficient

    matrix satisfy $\begin{vmatrix} 1-r & 0 & 0 \\ 2 & 1-r & -2 \\ 3 & 2 & 1-r \end{vmatrix} \begin{pmatrix} \xi_1 \\ \xi_2 \\ \xi_3 \end{pmatrix} = \begin{pmatrix} 0 \\ 0 \\ 0 \end{pmatrix}$. The

    determinant of coefficients reduces to $(1-r)(r^2 - 2r + 5)$
    so the eigenvalues are $r_1 = 1$, $r_2 = 1 + 2i$, and
    $r_3 = 1 - 2i$. The eigenvector corresponding to r_1
    satisfies

$$\begin{pmatrix} 0 & 0 & 0 \\ 2 & 0 & -2 \\ 3 & 2 & 0 \end{pmatrix} \begin{pmatrix} \xi_1 \\ \xi_2 \\ \xi_3 \end{pmatrix} = \begin{pmatrix} 0 \\ 0 \\ 0 \end{pmatrix}; \text{ hence } \xi_1 - \xi_3 = 0 \text{ and}$$

$(3\xi_1 + 2\xi_2 = 0.$ If we let $\xi_2 = -3$ then $\xi_1 = 2$ and $\xi_3 = 2$,

so one solution of the D.E. is $\begin{pmatrix} 2 \\ -3 \\ 2 \end{pmatrix} e^t$. The eigenvector

corresponding to r_2 satisfies $\begin{pmatrix} 2i & 0 & 0 \\ 2 & -2i & -2 \\ 3 & 2 & -2i \end{pmatrix} \begin{pmatrix} \xi_1 \\ \xi_2 \\ \xi_3 \end{pmatrix} = \begin{pmatrix} 0 \\ 0 \\ 0 \end{pmatrix}$.

Hence $\xi_1 = 0$ and $i\xi_2 + \xi_3 = 0$. If we let $\xi_2 = 1$, then
$\xi_3 = -i$. Thus a complex-valued solution is

$\begin{pmatrix} 0 \\ 1 \\ -i \end{pmatrix} e^t (\cos2t + i \sin2t)$. Taking the real and imaginary

parts we obtain $\begin{pmatrix} 0 \\ \cos2t \\ \sin2t \end{pmatrix} e^t$ and $\begin{pmatrix} 0 \\ \sin2t \\ -\cos2t \end{pmatrix} e^t$, respectively.

Thus the general solution is

$$\underset{\sim}{x} = c_1 \begin{pmatrix} 2 \\ -3 \\ 2 \end{pmatrix} e^t + c_2 e^t \begin{pmatrix} 0 \\ \cos2t \\ \sin2t \end{pmatrix} + c_3 e^t \begin{pmatrix} 0 \\ \sin2t \\ -\cos2t \end{pmatrix}.$$

9. The eigenvalues and eigenvectors of the coefficient

matrix satisfy $\begin{pmatrix} 1-r & -5 \\ 1 & -3-r \end{pmatrix} \begin{pmatrix} \xi_1 \\ \xi_2 \end{pmatrix} = \begin{pmatrix} 0 \\ 0 \end{pmatrix}$. The determinant of

coefficients is $r^2 + 2r + 2$ so that the eigenvalues are
$r = -1 \pm i$. The eigenvector corresponding to $r = -1 + i$

is given by $\begin{pmatrix} 2-i & -5 \\ 1 & -2-i \end{pmatrix} \begin{pmatrix} \xi_1 \\ \xi_2 \end{pmatrix} = \underset{\sim}{0}$ so that $\xi_1 = (2+i)\xi_2$ and

thus one complex-valued solution is

$\underset{\sim}{x}^{(1)}(t) = \begin{pmatrix} 2+i \\ 1 \end{pmatrix} e^{(-1+i)t}$. Finding the real and complex

parts of $\underset{\sim}{x}^{(1)}$ leads to the general solution

$$x = c_1 e^{-t} \begin{pmatrix} 2\cos t - \sin t \\ \cos t \end{pmatrix} + c_2 e^{-t} \begin{pmatrix} 2\sin t + \cos t \\ \sin t \end{pmatrix}. \quad \text{Setting}$$

$t = 0$ we find $x(0) = \begin{pmatrix} 1 \\ 1 \end{pmatrix} = c_1 \begin{pmatrix} 2 \\ 1 \end{pmatrix} + c_2 \begin{pmatrix} 1 \\ 0 \end{pmatrix}$, which is

equivalent to the system $\begin{array}{c} 2c_1 + c_2 = 1 \\ c_1 + 0 = 1 \end{array}$. Thus $c_1 = 1$ and

$c_2 = -1$ and

$$x(t) = e^{-t} \begin{pmatrix} 2\cos t - \sin t \\ \cos t \end{pmatrix} - e^{-t} \begin{pmatrix} 2\sin t + \cos t \\ \sin t \end{pmatrix}$$

$$= e^{-t} \begin{pmatrix} \cos t - 3\sin t \\ \cos t - \sin t \end{pmatrix}, \quad \text{which spirals to zero as}$$

$t \to \infty$ due to the e^{-t} term.

11. If we seek solutions of the form $x = \xi t^r$, then r must be
an eigenvalue and ξ a corresponding eigenvector of the
coefficient matrix. Thus r and ξ satisfy

$\begin{pmatrix} -1-r & -1 \\ 2 & -1-r \end{pmatrix} \begin{pmatrix} \xi_1 \\ \xi_2 \end{pmatrix} = \begin{pmatrix} 0 \\ 0 \end{pmatrix}$. The determinant of coefficients

is $(-1-r)^2 + 2 = r^2 + 2r + 3$, so the eigenvalues are
$r = -1 \pm \sqrt{2}\,i$. The eigenvector corresponding to

$-1 \pm \sqrt{2}\,i$ satisfies $\begin{pmatrix} -\sqrt{2}\,i & -1 \\ 2 & -\sqrt{2}\,i \end{pmatrix} \begin{pmatrix} \xi_1 \\ \xi_2 \end{pmatrix} = \begin{pmatrix} 0 \\ 0 \end{pmatrix}$ or

$\sqrt{2}\,i\xi_1 + \xi_2 = 0$. If we let $\xi_1 = 1$, then $\xi_2 = -\sqrt{2}\,i$, and

$\xi^{(1)} = \begin{pmatrix} 1 \\ -\sqrt{2}\,i \end{pmatrix}$. Thus a complex-valued solution of the

given D.E. is $\begin{pmatrix} 1 \\ -\sqrt{2}\,i \end{pmatrix} t^{-1+\sqrt{2}\,i}$. Referring to Section 5.5,

if necessary, we have
$t^{-1+\sqrt{2}\,i} = t^{-1}[\cos(\sqrt{2}\,\ln t) + i\sin(\sqrt{2}\,\ln t)]$ for $t > 0$.
Separating the complex valued solution into real and
imaginary parts, we obtain the two real-valued solutions

$$u = t^{-1} \begin{pmatrix} \cos(\sqrt{2}\,\ln t) \\ 2\sin(\sqrt{2}\,\ln t) \end{pmatrix} \text{ and } v = t^{-1} \begin{pmatrix} \sin(\sqrt{2}\,\ln t) \\ -2\cos(\sqrt{2}\,\ln t) \end{pmatrix}.$$

1. The eigenvalues and eigenvectors of the given coefficient
    matrix satisfy $\begin{pmatrix} 3-r & -4 \\ 1 & -1-r \end{pmatrix} \begin{pmatrix} \xi_1 \\ \xi_2 \end{pmatrix} = \begin{pmatrix} 0 \\ 0 \end{pmatrix}$. The determinant of
    coefficients is $(3-r)(-1-r) + 4 = r^2 - 2r + 1 = (r-1)^2$ so
    $r_1 = 1$ and $r_2 = 1$. The eigenvectors corresponding to
    this double eigenvalue satisfy $\begin{pmatrix} 2 & -4 \\ 1 & -2 \end{pmatrix} \begin{pmatrix} \xi_1 \\ \xi_2 \end{pmatrix} = \begin{pmatrix} 0 \\ 0 \end{pmatrix}$, or
    $\xi_1 - 2\xi_2 = 0$. Thus the only eigenvectors are multiples
    of $\xi^{(1)} = \begin{pmatrix} 2 \\ 1 \end{pmatrix}$. One solution of the given D.E. is
    $x^{(1)}(t) = \begin{pmatrix} 2 \\ 1 \end{pmatrix} e^t$, but there is no second solution of this
    form. To find a second solution we assume that
    $x = \eta t e^t + \zeta e^t$ and substitute this expression into the
    D.E. As in Example 1 we find that η is an eigenvector,
    so we choose $\eta = \begin{pmatrix} 2 \\ 1 \end{pmatrix}$. Then ζ must satisfy
    $\begin{pmatrix} 2 & -4 \\ 1 & -2 \end{pmatrix} \begin{pmatrix} \zeta_1 \\ \zeta_2 \end{pmatrix} = \begin{pmatrix} 2 \\ 1 \end{pmatrix}$, which verifies Eq.(20). Solving these
    equations yields $\zeta_1 - 2\zeta_2 = 1$. If $\zeta_2 = k$, where k is an
    arbitrary constant, then $\zeta_1 = 1 + 2k$. Hence the second
    solution that we obtain is
    $$x^{(2)}(t) = \begin{pmatrix} 2 \\ 1 \end{pmatrix} t e^t + \begin{pmatrix} 1 + 2k \\ k \end{pmatrix} e^t = \begin{pmatrix} 2 \\ 1 \end{pmatrix} t e^t + \begin{pmatrix} 1 \\ 0 \end{pmatrix} e^t + k \begin{pmatrix} 2 \\ 1 \end{pmatrix} e^t.$$
    The last term is a multiple of the first solution $x^{(1)}(t)$
    and may be neglected, that is, we may set $k = 0$. Thus
    $x^{(2)}(t) = \begin{pmatrix} 2 \\ 1 \end{pmatrix} t e^t + \begin{pmatrix} 1 \\ 0 \end{pmatrix} e^t$ and the general solution is
    $x = c_1 x^{(1)}(t) + c_2 x^{(2)}(t)$.

5. Substituting $x = \xi e^{rt}$ into the given system, we find that
    the eigenvalues and eigenvectors satisfy
    $\begin{pmatrix} 1-r & 1 & 1 \\ 2 & 1-r & -1 \\ 0 & -1 & 1-r \end{pmatrix} \begin{pmatrix} \xi_1 \\ \xi_2 \\ \xi_3 \end{pmatrix} = \begin{pmatrix} 0 \\ 0 \\ 0 \end{pmatrix}$. The determinant of coefficients

is $-r^3 + 3r^2 - 4$ and thus $r_1 = -1$, $r_2 = 2$ and $r_3 = 2$.
The eigenvector corresponding to r_1 satisfies

$$\begin{pmatrix} 2 & 1 & 1 \\ 2 & 2 & -1 \\ 0 & -1 & 2 \end{pmatrix} \begin{pmatrix} \xi_1 \\ \xi_2 \\ \xi_3 \end{pmatrix} = \begin{pmatrix} 0 \\ 0 \\ 0 \end{pmatrix} \text{ which yields } \xi^{(1)} = \begin{pmatrix} -3 \\ 4 \\ 2 \end{pmatrix} \text{ and}$$

$$x^{(1)} = \begin{pmatrix} -3 \\ 4 \\ 2 \end{pmatrix} e^{-t}.$$ The eigenvectors corresponding to the

double eigenvalue must satsify $\begin{pmatrix} -1 & 1 & 1 \\ 2 & -1 & -1 \\ 0 & -1 & -1 \end{pmatrix} \begin{pmatrix} \xi_1 \\ \xi_2 \\ \xi_3 \end{pmatrix} = \begin{pmatrix} 0 \\ 0 \\ 0 \end{pmatrix}$,

which yields the single eigenvector $\xi^{(2)} = \begin{pmatrix} 0 \\ 1 \\ -1 \end{pmatrix}$ and hence

$$x^{(2)}(t) = \begin{pmatrix} 0 \\ 1 \\ -1 \end{pmatrix} e^{2t}.$$ The second solution corresponding to

the double eigenvalue will have the form specified by

Eq. (19), which yields $x^{(3)} = \begin{pmatrix} 0 \\ 1 \\ -1 \end{pmatrix} te^{2t} + \eta e^{2t}$.

Substituting this into the given system, or using

Eq. (20), we find that η satisfies $\begin{pmatrix} -1 & 1 & 1 \\ 2 & -1 & -1 \\ 0 & -1 & -1 \end{pmatrix} \begin{pmatrix} \eta_1 \\ \eta_2 \\ \eta_3 \end{pmatrix} = \begin{pmatrix} 0 \\ 1 \\ -1 \end{pmatrix}$.

Using row reduction we find that $\eta_1 = 1$ and $\eta_2 + \eta_3 = 1$,
where either η_2 or η_3 is arbitrary. If we choose $\eta_2 = 0$,

then $\eta = \begin{pmatrix} 1 \\ 0 \\ 1 \end{pmatrix}$ and thus $x^{(3)} = \begin{pmatrix} 0 \\ 1 \\ -1 \end{pmatrix} te^{2t} + \begin{pmatrix} 1 \\ 0 \\ 1 \end{pmatrix} e^{2t}.$ The

general solution is then $x = c_1 x^{(1)} + c_2 x^{(2)} + c_3 x^{(3)}$.

9. The eigenvalues and eigenvectors of the coefficient

matrix satisfy $\begin{pmatrix} 1-r & 1 & 1 \\ 2 & 1-r & -1 \\ -3 & 2 & 4-r \end{pmatrix} \begin{pmatrix} \xi_1 \\ \xi_2 \\ \xi_3 \end{pmatrix} = \begin{pmatrix} 0 \\ 0 \\ 0 \end{pmatrix}$. The determinant

of coefficients is $8 - 12r + 6r^2 - r^3 = (2-r)^3$, so the
eigenvalues are $r_1 = r_2 = r_3 = 2$. The eigenvectors
corresponding to this triple eigenvalue satisfy

$$\begin{pmatrix} -1 & 1 & 1 \\ 2 & -1 & -1 \\ -3 & 2 & 2 \end{pmatrix} \begin{pmatrix} \xi_1 \\ \xi_2 \\ \xi_3 \end{pmatrix} = \begin{pmatrix} 0 \\ 0 \\ 0 \end{pmatrix}.$$ Using row reduction we can reduce

this to the equivalent system $\xi_1 - \xi_2 - \xi_3 = 0$, and
$\xi_2 + \xi_3 = 0$. If we let $\xi_2 = 1$, then $\xi_3 = -1$ and $\xi_1 = 0$,

so the only eigenvectors are multiples of $\underset{\sim}{\xi} = \begin{pmatrix} 0 \\ 1 \\ -1 \end{pmatrix}$.   Thus

one solution of the given D.E. is $\underset{\sim}{x}^{(1)}(t) = \begin{pmatrix} 0 \\ 1 \\ -1 \end{pmatrix} e^{2t}$, but

there are no other linearly independent solutions of this
form. We now seek a second solution of the form
$\underset{\sim}{x} = \underset{\sim}{\xi}te^{2t} + \underset{\sim}{\eta}e^{2t}$. As in the text, $\underset{\sim}{\xi}$ must be an

eigenvector we we choose $\underset{\sim}{\xi} = \begin{pmatrix} 0 \\ 1 \\ -1 \end{pmatrix}$. Then $\underset{\sim}{\eta}$ must satisfy a

system of the form given by Eq.(20, that is

$$\begin{pmatrix} -1 & 1 & 1 \\ 2 & -1 & -1 \\ -3 & 2 & 2 \end{pmatrix} \begin{pmatrix} \eta_1 \\ \eta_2 \\ \eta_3 \end{pmatrix} = \begin{pmatrix} 0 \\ 1 \\ -1 \end{pmatrix}.$$ By row reduction this is

equivalent to the system $\begin{pmatrix} 1 & -1 & -1 \\ 0 & 1 & 1 \\ 0 & 0 & 0 \end{pmatrix} \begin{pmatrix} \eta_1 \\ \eta_2 \\ \eta_3 \end{pmatrix} = \begin{pmatrix} 0 \\ 1 \\ 0 \end{pmatrix}.$ If we

choose $\eta_2 = 0$, then $\eta_3 = 1$ and $\eta_1 = 1$, so $\underset{\sim}{\eta} = \begin{pmatrix} 1 \\ 0 \\ 1 \end{pmatrix}$.   Hence

a second solution of the D.E. is

$$\underset{\sim}{x}^{(2)}(t) = \begin{pmatrix} 0 \\ 1 \\ -1 \end{pmatrix} te^{2t} + \begin{pmatrix} 1 \\ 0 \\ 1 \end{pmatrix} e^{2t}.$$ Finally, we seek a third

solution of the form $\underset{\sim}{x} = \underset{\sim}{\xi}(t^2/2)e^{2t} + \underset{\sim}{\eta}te^{2t} + \underset{\sim}{\zeta}e^{2t}$, from
Eq.(22), where $\underset{\sim}{\xi}$ and $\underset{\sim}{\eta}$ are as above, and $\underset{\sim}{\zeta}$ satisfies a

system of the form given by Eq.(23), namely

$$\begin{pmatrix} -1 & 1 & 1 \\ 2 & -1 & -1 \\ -3 & 2 & 2 \end{pmatrix} \begin{pmatrix} \zeta_1 \\ \zeta_2 \\ \zeta_3 \end{pmatrix} = \begin{pmatrix} 1 \\ 0 \\ 1 \end{pmatrix}.$$ By row reduction we find the

equivalent system $\begin{pmatrix} 1 & -1 & -1 \\ 0 & 1 & 1 \\ 0 & 0 & 0 \end{pmatrix} \begin{pmatrix} \zeta_1 \\ \zeta_2 \\ \zeta_3 \end{pmatrix} = \begin{pmatrix} -1 \\ 2 \\ 0 \end{pmatrix}.$ If we let

$\zeta_2 = 0$, then $\zeta_3 = 2$ and $\zeta_1 = 1$, so $\zeta = \begin{pmatrix} 1 \\ 0 \\ 2 \end{pmatrix}.$ Since any

multiple of a solution is again a solution it is
convenient to multiply by 2 and write

$$\underset{\sim}{x}^{(3)}(t) = \begin{pmatrix} 0 \\ 1 \\ -1 \end{pmatrix} t^2 e^{2t} + 2 \begin{pmatrix} 1 \\ 0 \\ 1 \end{pmatrix} te^{2t} + 2 \begin{pmatrix} 1 \\ 0 \\ 2 \end{pmatrix} e^{2t}.$$ The general

solution is then $\underset{\sim}{x} = c_1 \underset{\sim}{x}^{(1)}(t) + c_2 \underset{\sim}{x}^{(2)}(t) + c_3 \underset{\sim}{x}^{(3)}(t).$

11. Assuming $\underset{\sim}{x} = \underset{\sim}{\xi}t^r$ and substituting into the given system,

we find r and $\underset{\sim}{\xi}$ must satisfy $\begin{pmatrix} 1-r & -4 \\ 4 & -7-r \end{pmatrix} \begin{pmatrix} \xi_1 \\ \xi_2 \end{pmatrix} = \begin{pmatrix} 0 \\ 0 \end{pmatrix}$, which

has the double eigenvalue $r = -3$ and single eigenvector

$\begin{pmatrix} 1 \\ 1 \end{pmatrix}.$ Hence one solution of the given D.E. is

$\underset{\sim}{x}^{(1)}(t) = \begin{pmatrix} 1 \\ 1 \end{pmatrix} t^{-3}.$ By analogy with the scalar case

considered in Section 5.5 and Example 1 of this section,
we seek a second solution of the form $\underset{\sim}{x} = \underset{\sim}{\eta}t^{-3}\ln t + \underset{\sim}{\zeta}t^{-3}.$
Substituting this expression into the D.E. we find that $\underset{\sim}{\eta}$
and $\underset{\sim}{\zeta}$ satisfy the equations $(\underset{\sim}{A} - 3\underset{\sim}{I})\underset{\sim}{\eta} = \underset{\sim}{0}$ and

$(\underset{\sim}{A} - 3\underset{\sim}{I})\underset{\sim}{\zeta} = \underset{\sim}{\eta}$, where $\underset{\sim}{A} = \begin{pmatrix} 1 & -4 \\ 4 & -7 \end{pmatrix}$ and $\underset{\sim}{I}$ is the indentity

matrix. Thus $\underset{\sim}{\eta} = \begin{pmatrix} 1 \\ 1 \end{pmatrix},$ from above, and $\underset{\sim}{\zeta}$ is found to be

$\begin{pmatrix} 0 \\ -1/4 \end{pmatrix}.$ Thus a second solution is

$$\underset{\sim}{x}^{(2)}(t) = \begin{pmatrix} 1 \\ 1 \end{pmatrix} t^{-3} \ln t + \begin{pmatrix} 0 \\ -1/4 \end{pmatrix} t^3.$$

13. All solutions of the given system approach zero as $t \to \infty$ if and only if the eigenvalues of the coefficient matrix either are real and negative or else are complex with negative real part. Write down the determinantal equation satisfied by the eigenvalues and determine when the eigenvalues are as stated.

Section 7.8, Page 378

Each of the Problems 1 through 10 has been solved in one of the previous sections. Thus a fundamental matrix for the given systems can be readily written down. The fundamental matrix $\underset{\sim}{\Phi}(t)$ satisfying $\underset{\sim}{\Phi}(0) = \underset{\sim}{I}$ can then be found, as shown in the following problems.

4. From Problem 4 of Section 7.5 we have the two linearly independent solutions $\underset{\sim}{x}^{(1)}(t) = \begin{pmatrix} 1 \\ -4 \end{pmatrix} e^{-3t}$ and

$\underset{\sim}{x}^{(2)}(t) = \begin{pmatrix} 1 \\ 1 \end{pmatrix} e^{2t}$. Hence a fundamental matrix $\underset{\sim}{\psi}$ is given

by $\underset{\sim}{\Psi}(t) = \begin{pmatrix} e^{-3t} & e^{2t} \\ -4e^{-3t} & e^{2t} \end{pmatrix}$. To find the fundamental matrix

$\underset{\sim}{\Phi}(t)$ satisfying the I.C. $\underset{\sim}{\Phi}(0) = \underset{\sim}{I}$ we can proceed in either of two ways. One way is to find $\underset{\sim}{\Psi}(0)$, invert it to obtain $\underset{\sim}{\Psi}^{-1}(0)$, and then to form the product $\underset{\sim}{\Psi}(t)\underset{\sim}{\Psi}^{-1}(0)$, which is $\underset{\sim}{\Phi}(t)$. Alternatively, we can find the first column of $\underset{\sim}{\Phi}$ by determining the liner combination

$c_1\underset{\sim}{x}^{(1)}(t) + c_2\underset{\sim}{x}^{(2)}(t)$ that satisfies the I.C. $\begin{pmatrix} 1 \\ 0 \end{pmatrix}$. This

requires that $c_1 + c_2 = 1$, $-4c_1 + c_2 = 0$, so we obtain $c_1 = 1/5$ and $c_2 = 4/5$. Thus the first column of $\underset{\sim}{\Phi}(t)$ is

$\begin{pmatrix} (1/5)e^{-3t} + (4/5)e^{2t} \\ -(4/5)e^{-3t} + (4/5)e^{2t} \end{pmatrix}$. Similarly, the second column of

$\underset{\sim}{\Phi}$ is that linear combination of $\underset{\sim}{x}^{(1)}(t)$ and $\underset{\sim}{x}^{(2)}(t)$ that satisfies the I.C. $\begin{pmatrix} 0 \\ 1 \end{pmatrix}$. Thus we must have

$c_1 + c_2 = 0$, $-4c_1 + c_2 = 1$; therefore $c_1 = -1/5$ and $c_2 = 1/5$. Hence the second column of $\underset{\sim}{\Phi}(t)$ is

$$\begin{pmatrix} -(1/5)e^{-3t} + (1/5)e^{2t} \\ (4/5)e^{-3t} + (1/5)e^{2t} \end{pmatrix}.$$

6. Two linearly independent real-valued solutions of the given D.E. were found in Problem 2 of Section 7.6. Using the result of that problem, we have

$\underset{\sim}{\Psi}(t) = \begin{pmatrix} -2e^{-t}\sin2t & 2e^{-t}\cos2t \\ e^{-t}\cos2t & e^{-t}\sin2t \end{pmatrix}$. To find $\underset{\sim}{\Phi}(t)$

we determine the linear combinations of the columns of $\underset{\sim}{\Psi}(t)$ that satisfy the I.C. $\begin{pmatrix} 1 \\ 0 \end{pmatrix}$ and $\begin{pmatrix} 0 \\ 1 \end{pmatrix}$, respectively.

In the first case c_1 and c_2 satisfy $0c_1 + 2c_2 = 1$ and $c_1 + 0c_2 = 0$. Thus $c_1 = 0$ and $c_2 = 1/2$. In the second case we have $0c_1 + 2c_2 = 0$ and $c_1 + 0c_2 = 1$, so $c_1 = 1$ and $c_2 = 0$. Using these values of c_1 and c_2 to form the first and second columns of $\underset{\sim}{\Phi}(t)$ respectively, we obtain

$$\underset{\sim}{\Phi}(t) = \begin{pmatrix} e^t\cos2t & -2e^{-t}\sin2t \\ (1/2)e^{-t}\sin2t & e^{-t}\cos2t \end{pmatrix}.$$

8. Two linearly independent solutions of this D.E. were found in Problem 1 of Section 7.7. Using that result we

have $\underset{\sim}{\Psi}(t) = \begin{pmatrix} 2e^t & e^t + 2te^t \\ e^t & te^t \end{pmatrix}$. To find $\underset{\sim}{\Phi}(t)$ we determine

the linear combinations of the columns of $\underset{\sim}{\Psi}(t)$ that

satisfy the I.C. $\begin{pmatrix} 1 \\ 0 \end{pmatrix}$ and $\begin{pmatrix} 0 \\ 1 \end{pmatrix}$, respectively. In the first

case we have $2c_1 + c_2 = 1$ and $c_1 + 0c_2 = 0$, so $c_1 = 0$ and $c_2 = 1$. In the second case $2c_1 + c_2 = 0$ and $c_1 + 0c_2 = 1$, so $c_1 = 1$ and $c_2 = -2$. Using these values of c_1 and c_2 to form the respective columns of $\underset{\sim}{\Phi}(t)$ we

obtain $\underset{\sim}{\Phi}(t) = \begin{pmatrix} e^t + 2te^t & -4te^t \\ te^t & e^t - 2te^t \end{pmatrix}.$

14a. $A^2 = AA = \begin{pmatrix} \lambda & 1 \\ 0 & \lambda \end{pmatrix}\begin{pmatrix} \lambda & 1 \\ 0 & \lambda \end{pmatrix} = \begin{pmatrix} \lambda^2 & 2\lambda \\ 0 & \lambda^2 \end{pmatrix}$

$A^3 = AA^2 = \begin{pmatrix} \lambda & 1 \\ 0 & \lambda \end{pmatrix}\begin{pmatrix} \lambda^2 & 2\lambda \\ 0 & \lambda^2 \end{pmatrix} = \begin{pmatrix} \lambda^3 & 3\lambda \\ 0 & \lambda^3 \end{pmatrix}$

14b. Based upon the results of part a, assume

$A^n = \begin{pmatrix} \lambda^n & n\lambda^{n-1} \\ 0 & \lambda^n \end{pmatrix}$, then

$A^{n+1} = AA^n = \begin{pmatrix} \lambda & 1 \\ 0 & \lambda \end{pmatrix}\begin{pmatrix} \lambda^n & n\lambda^{n-1} \\ 0 & \lambda^n \end{pmatrix} = \begin{pmatrix} \lambda^{n+1} & (n+1)\lambda^n \\ 0 & \lambda^{n+1} \end{pmatrix}$, which is

the same as A^n with n replaced by n+1. Thus, by mathematical induction, A^n has the form shown above.

14c. From Eq.(29) we have

$\exp(At) = I + \sum_{n=1}^{\infty} \frac{A^n t^n}{n!}$

$= I + \sum_{n=1}^{\infty} \begin{pmatrix} \dfrac{\lambda^n t^n}{n!} & \dfrac{n\lambda^{n-1} t^n}{n!} \\ 0 & \dfrac{\lambda^n t^n}{n!} \end{pmatrix}$

$= \begin{pmatrix} 1 + \sum_{n=1}^{\infty} \dfrac{\lambda^n t^n}{n!} & \sum_{n=1}^{\infty} \dfrac{\lambda^{n-1} t^n}{(n-1)!} \\ 0 & 1 + \sum_{n=1}^{\infty} \dfrac{\lambda^n t^n}{n!} \end{pmatrix}$

$= \begin{pmatrix} e^{\lambda t} & te^{\lambda t} \\ 0 & e^{\lambda t} \end{pmatrix}$, since

$\sum_{n=1}^{\infty} \frac{\lambda^{n-1} t^n}{(n-1)!} = t(1 + \sum_{n=1}^{\infty} \frac{\lambda^n t^n}{n!}) = te^{\lambda t}$.

Section 7.9, Page 385

1. From Section 7.5 Problem 3 we have

$\underset{\sim}{x}^{(c)} = c_1 \begin{pmatrix} 1 \\ 1 \end{pmatrix} e^t + c_2 \begin{pmatrix} 1 \\ 3 \end{pmatrix} e^{-t}$. Note that

$\underset{\sim}{g}(t) = \begin{pmatrix} 1 \\ 0 \end{pmatrix} e^t + \begin{pmatrix} 0 \\ 1 \end{pmatrix} t$ and that r = 1 is an eigenvalue of the coefficient matrix. Thus if the method of undetermined coefficients is used, the assumed form is given by Eq.(18).

2. Using methods of previous sections, we find that the eigenvalues are $r_1 = 2$ and $r_2 = -2$, with corresponding eigenvectors $\begin{pmatrix} \sqrt{3} \\ 1 \end{pmatrix}$ and $\begin{pmatrix} 1 \\ -\sqrt{3} \end{pmatrix}$. Thus

$$\underset{\sim}{x}^{(c)} = c_1 \begin{pmatrix} \sqrt{3} \\ 1 \end{pmatrix} e^{2t} + c_2 \begin{pmatrix} 1 \\ -\sqrt{3} \end{pmatrix} e^{-2t}.$$ Writing the

nonhomogeneous term as $\begin{pmatrix} 1 \\ 0 \end{pmatrix} e^t + \begin{pmatrix} 0 \\ \sqrt{3} \end{pmatrix} e^{-t}$ we see that we

can assume $\underset{\sim}{x}^{(p)} = \underset{\sim}{a} e^t + \underset{\sim}{b} e^{-t}$. Substituting this in the D.E., we obtain

$$\underset{\sim}{a} e^t - \underset{\sim}{b} e^{-t} = \underset{\sim\sim}{A} \underset{\sim}{a} e^t + \underset{\sim\sim}{A} \underset{\sim}{b} e^{-t} + \begin{pmatrix} 1 \\ 0 \end{pmatrix} e^t + \begin{pmatrix} 0 \\ \sqrt{3} \end{pmatrix} e^{-t},$$ where $\underset{\sim}{A}$ is

the given coefficient matrix. All the terms involving e^t must add to zero and thus we have $\underset{\sim\sim}{A} \underset{\sim}{a} - \underset{\sim}{a} + \begin{pmatrix} 1 \\ 0 \end{pmatrix} = \begin{pmatrix} 0 \\ 0 \end{pmatrix}$.

This is equivalent to the system $\sqrt{3} a_2 = -1$ and $\sqrt{3} a_1 - 2a_2 = 0$, or $a_1 = -2/3$ and $a_2 = -1/\sqrt{3}$. Likewise the terms involving e^{-t} must add to zero, which yields $\underset{\sim\sim}{A} \underset{\sim}{b} + \underset{\sim}{b} + \begin{pmatrix} 0 \\ \sqrt{3} \end{pmatrix} = \begin{pmatrix} 0 \\ 0 \end{pmatrix}$. The solution of this system is $b_1 = -1$ and $b_2 = 2/\sqrt{3}$. Substituting these values for $\underset{\sim}{a}$ and $\underset{\sim}{b}$ into $\underset{\sim}{x}^{(p)}$ and adding $\underset{\sim}{x}^{(p)}$ to $\underset{\sim}{x}^{(c)}$ yields the desired solution.

3. The method of undetermined coefficients is not straight forward since the assumed form of $\underset{\sim}{x}^{(p)} = \underset{\sim}{a} \cos t + \underset{\sim}{b} \sin t$ leads to singular equations for $\underset{\sim}{a}$ and $\underset{\sim}{b}$. From Problem 3 of Section 7.6 we find that a fundamental matrix is
$$\underset{\sim}{\Psi}(t) = \begin{pmatrix} 5\cos t & 5\sin t \\ 2\cos t + \sin t & -\cos t + 2\sin t \end{pmatrix}.$$ The inverse
matrix is found (using steps similar to those illustrated in Example 2 of Section 7.2) to be

$$\Psi^{-1}(t) = \begin{pmatrix} \dfrac{\cos t - 2\sin t}{5} & \sin t \\ \dfrac{2\cos t + \sin t}{5} & -\cos t \end{pmatrix}. \quad \text{Thus we may use the}$$

method of variation of parameters where $x = \Psi(t)u(t)$ and $u(t)$ is given by $u'(t) = \Psi^{-1}(t)g(t)$ from Eq.(27). For

this problem $g(t) = \begin{pmatrix} -\cos t \\ \sin t \end{pmatrix}$ and thus

$$u'(t) = \begin{pmatrix} \dfrac{\cos t - 2\sin t}{5} & \sin t \\ \dfrac{2\cos t + \sin t}{5} & -\cos t \end{pmatrix} \begin{pmatrix} -\cos t \\ \sin t \end{pmatrix}$$

$$= \frac{1}{5}\begin{pmatrix} 2 - 3\cos 2t + \sin 2t \\ -1 - \cos 2t - 3\sin 2t \end{pmatrix},$$

after multiplying and using appropriate trigonometric identities. Integration and multiplication by Ψ yields the desired solution.

4. In this problem we use the method illustrated in Example 1. From Problem 4 of Section 7.5 we have the

transformation matrix $T = \begin{pmatrix} 1 & 1 \\ -4 & 1 \end{pmatrix}$. Inverting T we find

that $T^{-1} = \dfrac{1}{5}\begin{pmatrix} 1 & -1 \\ 4 & 1 \end{pmatrix}$. If we let $x = Ty$ and substitute

into the D.E., we obtain

$$y' = \frac{1}{5}\begin{pmatrix} 1 & -1 \\ 4 & 1 \end{pmatrix}\begin{pmatrix} 1 & 1 \\ 4 & -2 \end{pmatrix}\begin{pmatrix} 1 & 1 \\ -4 & 1 \end{pmatrix}y + \frac{1}{5}\begin{pmatrix} 1 & -1 \\ 4 & 1 \end{pmatrix}\begin{pmatrix} e^{-2t} \\ -2e^t \end{pmatrix}$$

$$= \begin{pmatrix} -3 & 0 \\ 0 & 2 \end{pmatrix}y + \frac{1}{5}\begin{pmatrix} e^{-2t} + 2e^t \\ 4e^{-2t} - 2e^t \end{pmatrix}. \quad \text{This corresponds to}$$

the two scalar equations
$$y_1' + 3y_1 = (1/5)e^{-2t} + (2/5)e^t,$$
$$y_2' - 2y_2 = (4/5)e^{-2t} - (2/5)e^t,$$
which may be solved by the methods of Section 2.1. For the first equation the integrating factor is e^{3t} and we obtain $(e^{3t}y_1)' = (1/5)e^t + (2/5)e^{4t}$, so $e^{3t}y_1 = (1/5)e^t + (1/10)e^{4t} + c_1$. For the second equation the integrating factor is e^{-2t}, so $(e^{-2t}y_2)' = (4/5)e^{-4t} - (2/5)e^{-t}$. Hence

$e^{-2t}y_2 = -(1/5)e^{-4t} + (2/5)e^{-t} + c_2.$ Thus

$$\underset{\sim}{y} = \begin{pmatrix} 1/5 \\ -1/5 \end{pmatrix} e^{-2t} + \begin{pmatrix} 1/10 \\ 2/5 \end{pmatrix} e^t + \begin{pmatrix} c_1 e^{-3t} \\ c_2 e^{2t} \end{pmatrix}.$$ Finally,

multiplying by $\underset{\sim}{T}$, we obtain

$$\underset{\sim}{x} = \underset{\sim}{T}\underset{\sim}{y} = \begin{pmatrix} 0 \\ -1 \end{pmatrix} e^{-2t} + \begin{pmatrix} 1/2 \\ 0 \end{pmatrix} e^t + c_1 \begin{pmatrix} 1 \\ -4 \end{pmatrix} e^{-3t} + c_2 \begin{pmatrix} 1 \\ 1 \end{pmatrix} e^{2t}.$$

The last two terms are the general solution of the
correspoinding homogeneous system, while the first two
terms are a particular solution of the nonhomogeneous
system.

12. Since the coefficient is the same as that of Problem 3,
    use the same procedure as done in that problem, including
    the Ψ^{-1} found there. In the interval $\pi/2 < t < \pi$ note
    that $\sin t > 0$ and $\cos t < 0$; hence $|\sin t| = \sin t$, but
    $|\cos t| = -\cos t$.

14. To verify that the given vector is the general solution
    of the corresponding system, it is sufficient to
    substitute it into the D.E. Note also that the two terms
    in $\underset{\sim}{x}^{(c)}$ are linearly independent. If we seek a solution
    of the form $\underset{\sim}{x} = \Psi(t)\underset{\sim}{u}(t)$ then we find that the equation
    corresponding to Eq.(26) is $t\Psi(t)\underset{\sim}{u}'(t) = \underset{\sim}{g}(t)$, where

$$\Psi(t) = \begin{pmatrix} t & 1/t \\ t & 3/t \end{pmatrix} \text{ and } \underset{\sim}{g}(t) = \begin{pmatrix} 1-t^2 \\ 2t \end{pmatrix}.$$ Thus

$\underset{\sim}{u}' = (1/t)\Psi^{-1}(t)\underset{\sim}{g}(t).$ Using row operations on Ψ and I,

we find that $\Psi^{-1} = \begin{pmatrix} 3/2t & -1/2t \\ -t/2 & t/2 \end{pmatrix}$ and thus

$$\underset{\sim}{u}' = \begin{pmatrix} 3/2t^2 & -3/2 & -1/t \\ -1/2 & + t^2/2 + t \end{pmatrix}.$$ Integration and multiplication

by $\Psi(t)$ yields the desired solution.

134

Section 8.1, Page 392

2a. The Euler formula is $y_{n+1} = y_n + h(2y_n - t_n + 1/2)$ for
n = 0,1,2,3 and with $t_0 = 0$ and $y_0 = 1$. Thus
$y_1 = y_0 + .1(2y_0 - t_0 + 1/2) = 1.25$,
$y_2 = 1.25 + .1[2(1.25) - (.1) + 1/2] = 1.54$,
$y_3 = 1.54 + .1[2(1.54) - (.2) + 1/2] = 1.878$, and
$y_4 = 1.878 + .1[2(1.878) - (.3) + 1/2] = 2.2736$.

2b. Use the same formula as in Problem 2a, except now h = .05
and n = 0,1...7. Notice that only results for n = 1,3,5
and y are needed to compare with part a.

2c. Again, use the same formula as above with h = .025 and n
= 0,1...15. Notice that only results for n = 3,7,11 and
15 are needed to compare with parts a and b.

2d. $y' = 1/2 - t + 2y$ is a first order linear D.E. Rewrite
the equation in the form $y' - 2y = 1/2 - t$ and multiply
both sides by the integrating factor e^{-2t} to obtain
$(e^{-2t}y)' = (1/2 - t)e^{-2t}$. Integrating the right side by
parts and multiplying by e^{2t} we obtain $y = ce^{-2t} + t/2$.
The I.C. $y(0) = 1 \rightarrow c = 1$ and hence the solution of the
I.V.P. is $y = \phi(x) = e^{2t} + t/2$. Thus $\phi(0.1) = 1.2714$,
$\phi(0.2) = 1.59182$, $\phi(0.3) = 1.97212$, and $\phi(0.4) = 2.42554$.

5. The Euler formual is $y_{n+1} = y_n + h\sqrt{t_n + y_n}$ for
n = 0,1,2... with $t_0 = 1$ and $y_0 = 3$.

7. $y_1 = y_0 + .1\dfrac{y_0^2 + 2t_0 y_0}{3 + t_0^2} = 1 + .1\dfrac{4 + 4}{3 + 1} = 2.2$

$y_2 = y_1 + .1\dfrac{y_1^2 + 2t_1 y_1}{3 + t_1^2}$

$= 2.2 + .1\dfrac{(2.2)^2 + 2(1.1)(2.2)}{3 + (1.1)^2} = 2.42993$.

9. For part a forty steps must be taken, that is,
n = 0,1,...39 and for part b eighty steps must taken with
n = 0,1,...79. Thus use of a programmable calculator or
a computer is desirable.

13. Use the hint along with the values of y that you
calculate for t between 1.6 and 1.8. Note that the slope
(y') is positive for values of y < 1.155.

15. If $y' = 1 - t + 4y$ then
     $y'' = -1 + 4y' = -1 + 4(1-t+4y) = 3 - 4t + 16y$. In
     Eq.(12) we let y_n, y_n' and y_n'' denote the approximate
     values of $\phi(t_n)$, $\phi'(t_n)$, and $\phi''(t_n)$, respectively.
     Keeping the first three terms in the Taylor series we
     have
     $y_{n+1} = y_n + y_n'h + y_n'' h^2/2$
     $\qquad = y_n + (1 - t_n + 4y_n)h + (3 - 4t_n + 16y_n)h^2/2$ for
     $n = 0,1$ with $t_0 = 0$ and $y_0 = 1$.

17b. Using Eq.(6) we have $y_{n+1} = y_n + h(2y_n - 1) = (1+2h)y_n - h$.
     Setting $n + 1 = k$ (and hence $n = k = -1$) this becomes
     $y_k = (1 + 2h)y_{k-1} - h$, for $k = 1,2,\ldots$. Since $y_0 = 1$,
     we have $y_1 = 1 + 2h - h = 1 + h = (1 + 2h)/2 + 1/2$, and
     hence $y_2 = (1 + 2h)y_1 - h = (1 + 2h)^2/2 + (1 + 2h)/2 - h$
     $\qquad = (1 + 2h)^2/2 + 1/2$;
     $y_3 = (1 + 2h)y_2 - h = (1 + 2h)^3/2 + (1 + 2h)/2 - h$
     $\quad = (1 + 2h)^3/2 + 1/2$. Continuing in this fashion (or
     using induction) we obtain $y_k = (1 + 2h)^k/2 + 1/2$. For
     fixed $x > 0$ choose $h = x/k$. Then substitute for h in the
     last formula to obtain $y_k = (1 + 2x/k)^k/2 + 1/2$. Letting
     $k \to \infty$ we find $y(x) = y_k \to e^{2x}/2 + 1/2$, which is the
     exact solution. (See hint for Problem 16d.)

18a. $\phi_{n+1}(t) = 1 + \int_0^t [2\phi_n(s) - 1]ds$, $\phi_0(t) = 1$

     $\phi_1(t) = 1 + \int_0^t (2-1)ds = 1 + t$

     $\phi_2(t) = 1 + \int_0^t [2(1+s) - 1]ds = 1 + t + t^2$

     $\phi_3(t) = 1 + \int_0^t [2(1+s+s^2) - 1]ds = 1 + t + t^2 + \frac{2}{3}t^3$.

Section 8.2, Page 398

1.   If $y = \phi(t)$ is the exact solution of the I.V.P., then
     $\phi'(t) = 2\phi(t) - 1$ and $\phi''(t) = 2\phi'(t) = 4\phi(t) - 2$. From
     Eq.(10), $e_{n+1} = [2\phi(\bar{t}_n) - 1]h^2$, $t_n < \bar{t}_n < t_n + h$. Thus
     $|e_{n+1}| \le [1 + 2\max_{0 \le t \le 1} |\phi(t)|]h^2$. Since the exact solution
     is $y = \phi(t) = [1 + \exp(2t)]/2$, $e_{n+1} = h^2\exp(2\bar{t}_n)$.
     Therefore $|e_1| \le (0.1)^2\exp(0.2) = 0.012$ and
     $|e_4| \le (0.1)^2\exp(0.8) = 0.022$, since the maximum value of
     $\exp(2\bar{t}_n)$ occurs at $t = .1$ and $t = .4$ respectively.

4. The local formula error is $e_{n+1} = \phi''(\bar{t}_n)h^2/2$. For this
    problem $\phi'(t) = 5t - 3\phi^{1/2}(t)$ and thus
    $\phi''(t) = 5 - (3/2)\phi^{-1/2}\phi' = 19/2 - (15/2)t\phi^{-1/2}$.
    Substituting this last expression into e_{n+1} yields the
    desired answer.

7d. $e_{n+1} = -(5\pi/2)\sin(5\pi\bar{t}_n)h^2$.

8a. From Eq.(3) we have $E_n = \phi(t_n) - y_n$. Using this in
    Eq.(9) we obtain

    $E_{n+1} = E_n + h\{f[t_n, \phi(t_n)] - f(t_n,y_n)\} + \phi''(\bar{t}_n)h^2/2$. Using
    the given inequality involving L we have
    $|f[t_n, \phi(t_n)] - f(t_n,y_n)| \leq L |\phi(t_n) - y_n| = L|E_n|$ and
    thus
    $|E_{n+1}| \leq |E_n| + hL|E_n| + \max_{t_0 \leq t \leq t_n} |\phi''(t)|h^2/2 = \alpha|E_n| + \beta h^2$.

8b. Since $\alpha = 1 + hL$, $\alpha - 1 = hL$. Hence $\beta h^2(\alpha^n-1)/(\alpha-1) =$
    $\beta h^2[(1+hL)^n - 1]hL = \beta h[(1+hL)^n - 1]/L$.

8c. $(1+hL)^n \leq \exp(nhL)$ follows from the observation that
    $\exp(nhL) = [\exp(hL)]^n = (1 + hL + h^2L^2/2! + ...)^n$.
    Noting that $nh = x_n - x_0$, the rest follows from Eq.(ii).

10b. Using a step size of .2 we estimate $\phi(.2) = 1.2$ in one
    step. Using a step size of .1 we estimate $\phi(.2) = 1.22$
    in two steps. An improved estimate of $\phi(.2)$ is now given
    by Eq.(i) of Problem 9, which yields
    $\phi(.2) = 1.22 + (1.22 - 1.2) = 1.24$.

12. Using the result of Problem 9 and the entries from Table
    8.1.1 we have
    a. $\phi(1) \cong 45.588400 + 11.176910 = 56.765310$
    b. $\phi(1) \cong 53.807866 + 8.219466 = 62.027332$
    Although the proof suggested in Problem 11 is not
    applicable in this case, it is possible to show that the
    result is still valid using more advanced analysis.
    Thus, since the result of part (a) gives an estimate for
    $\phi(1)$ with an error proportional to h^2 with a step size of
    .05 and the result of part (b) gives an estimate for $\phi(1)$
    with an error proportional to h^2 with a step size of .025
    we may obtain a new estimate for $\phi(1)$ with an error
    proportional to h^3 with a step size of .025 as follows:
    $\phi(1) \cong 62.027332 + (62.027332 - 56.765310)/(2^2-1)$
         $= 63.781339$.

Section 8.3, Page 404

1a. The improved Euler formula is
$y_{n+1} = y_n + [y_n' + f(t_n + h, y_n + hy_n')]h/2$ where
$y' = f(t,y) = 2y - 1$. Hence $y_n' = 2y_n-1$, and
$f(t_n + h, y_n + hy_n') = 2(y_n + hy_n') - 1$. Thus we obtain
$y_{n+1} = y_n + [y_n' + 2(y_n + hy_n') -1]h/2$. If desired, we can
substitute for y_n' in terms of y_n and obtain
$y_{n+1} = y_n + h(1+h)(2y_n-1)$, $n = 0,1,2,3$ with $y_0 = 1$. In
this case the formula for y_{n+1} was made simpler by
substituting for y_n'; but this may not always be true.
Thus $y_1 = 1 + .1(1.1)(1) = 1.11$ and
$y_2 = 1.11 + .1(1.1)(1.22) = 1.2442$.

1b. If $y' = 2y - 1$ then $y'' = 2y' = 2(2y-1)$. Hence, the three-
term Taylor series formula is $y_{n+1} = y_n + y_n'h + y_n''h^2/2 =$
$y_n + y_n'h + y_n'h^2$ with $y_n' = 2y_n -1$. If desired, we can
substitute for y_n' and obtain $y_{n+1} = y_n + h(1+h)(2y_n-1)$,
$n = 0,1,2$ with $y_0 = 1$. Thus $y_1 = 1 + .1(1.1)(1) = 1.11$
and $y_2 = 1.11 + .1(1.1)(1.22) = 1.2442$. This problem
illustrates the result that the improved Euler and three-
term Taylor formulas are identical when f (=2y-1) is
linear in t and y.

4b. Since $y' = 5t -3\sqrt{y}$, we have $y'' = 5 - (3/2)y'/\sqrt{y} =$
$(19 - 15ty^{-1/2})/2$. Hence the three-term Taylor formula
is $y_{n+1} = y_n + h(5t_n - 3y_n^{1/2}) + h^2(19-15t_ny_n^{-1/2})/4$. Thus
$y_1 = 2 + .1(-3\sqrt{2}) + .01(19)/4 = 1.62324$ and
$y_2 = 1.62324 + .1(.5-3\sqrt{1.62324} + .01[19-15(.1)/\sqrt{1.62324}]/4$
$= 1.33557$.

5a. The improved Euler formula is $y_{n+1} = y_n + [y_n' + f(t_n + h,$
$y_n = hy_n')]h/2$ where $y' = f(t,y) = \sqrt{t+y}$. Hence
$y_n' = \sqrt{t_n+y_n}$ and
$f(t_n+h, y_n + hy_n') = \sqrt{(t_n+h)+(y_n+h\sqrt{t_n+y_n})}$. Thus we
obtain

$$y_1 = 3 + \frac{\sqrt{1+3} + \sqrt{(1.1)+(3+.1\sqrt{1+3})}}{2}(.1) = 3.20368 \text{ and}$$

$$y_2 = \frac{\sqrt{1.1+3.20368} + \sqrt{(1.2)+(3.20368 + .1\sqrt{1.1+3.20368})}}{2}(.1)$$
$$= 3.41478.$$

For Problems 9 through 12, since 40 steps are needed for part
(c), use a programmable calculator or a computer program
utilizing the BASIC program (or comparable program) given in
the text.

14. Since $f(t,y)$ is linear in t and y, the results should be
the same as those for the improved Euler method.

15a. Since $\phi(t_n + h) = \phi(t_{n+1})$ we have, using Eq.(5) and the
given equation, $e_{n+1} = \phi(t_{n+1}) - y_{n+1} + [\phi(t_n) - y_n] +$

$[\phi'(t_n) - \dfrac{y'_n + f(t_n+h, \ y_n+hy'_n)}{2}]h + \phi''(t_n)h^2/2! +$

$\phi'''(\bar{t}_n)h^3/3!$. Since $y_n = \phi(t_n)$ and $y'_n = \phi'(t_n) = f(t_n, y_n)$
this reduces to $e_{n+1} + \phi''(t_n)h^2/2! - \{f[t_n+h, \ y_n +$

$hf(t_n, y_n)] - f(t_n, y_n)\}h/2! + \phi'''(\bar{t}_n)h^3/3!$, which can be
written in the form of Eq.(i).

15b. First observe that $y' = f(t,y)$ and $y'' = f_t(t,y) +$
$f_y(t,y)y'$. Hence $\phi''(t_n) = f_t(t_n, y_n) + f_y(t_n, y_n)f(t_n, y_n)$.
Using the given Taylor series, with $a = t_n$, $h = h$, $b = y_n$
and $k = hf(t_n, y_n)$ we have
$f[t_n+h, y_n+hf(t_n, y_n)] = f(t_n, y_n) + f_t(t_n, y_n)h + f_y(t_n, y_n)hf(t_n, y_n)$
$+ [f_{tt}(\xi, \eta)h^2 + 2f_{ty}(\xi, \eta)h^2 f(t_n, y_n) + f_{yy}(\xi, \eta)h^2 f^2(t_n, y_n)]/2!$
where $t_n < \xi < t_n + h$ and $|\eta - y_n| < h|f(t_n, y_n)|$.
Substituting this in Eq.(i) and using the earlier
expression for $\phi''(t_n)$ we find that the first term on the
right side of Eq.(i) reduces to
$-[f_{tt}(\xi, \eta) + 2f_{ty}(\xi, \eta)f(t_n, y_n) + f_{yy}(\xi, \eta)f^2(t_n, y_n)]h^3/4$,
which is proportional to h^3 plus, possibly, higher order
terms. The reason that there may be higher order terms
is because ξ and η will, in general, depend upon h.

17. Since $\phi(t) = [4t - 3 + 19\exp(4t)]/16$ we have
$\phi'''(t) = 76\exp(4t)$ and thus from Problem 15c we find
$e_{n+1} = 38[\exp(4\bar{t}_n)]h^3/3$. Thus
$|e_{n+1}| \le (38h^3/3)\exp(4) = 691.6h^3$ on $0 \le t \le 1$.
$|e_1| = |\phi(t_1) - y_1| \le (0.038/3)\exp(0.4) = 0.0188964$,
which is approximately 1/10 of the error indicated in
Eq.(15) of the previous section.

23. The modified Euler formula is
$y_{n+1} = y_n + hf[t_nh/2, \ y_n + (h/2)f(t_n, y_n)]$ where
$f(t,y) = t^2 + y^2$. Since $t_0 = 0$, $y_0 = 1$ and $h = .1$, we

have $y_1 = 1 + .1[(.05)^2 + (1+.05)^2] = 1.1105$ and
$y_2 = 1.1105 + .1\{(.15)^2 + [1.1105+.05(.1^2+1.1105^2)]^2\}$
   $= 1.25026$.

26. Since $\phi(t_n,+h) = \phi(t_{n+1})$, we have, utilizing Eq.(10) and
    Eq.(13), $e_{n+1} = \phi(t_{n+1}) - y_{n+1} = \phi(t_n) + \phi'(t_n)h +$

    $\phi''(t_n)h^2/2! + \phi'''(\bar{t}_n)h^3/3! - y_n - hy_n' - h^2y_n''/2 =$

    $\phi'''(\bar{t}_n)h^3/3!$ where $t_n \le \bar{t}_n \le t_n + h$. Note that we have
    made use of the fact that $y_n = \phi(t_n)$, which means that
    $y_n' = f(t_n,y_n) = f[t_n,\phi(t_n)] = \phi'(t_n)$ and from Eq.(14)
    $y_n'' = f_t(t_n,y_n) + f_y(t_n,y_n)y_n' = f_t[t_n,\phi(t_n)] +$
    $f_y[t_n,\phi(t_n)]\phi'(t_n) = \phi''(t_n)$.

27. The four-term Taylor series formula for $y' = f(t,y)$ is
    $y_{n+1} = y_n + y_n'h + y_n''h^2/2! + y_n'''h^3/3!$, where $y_n' = f(t_n,y_n)$,
    $y_n'' = f_t(t_n,y_n) + f_y(t_n,y_n)y_n'$, and $y_n''' = f_{tt}(t_n,y_n) +$
    $f_{ty}(t_n,y_n)y_n' + [f_{yt}(t_n,y_n) + f_{yy}(t_n,y_n)y_n']y_n' +$
    $f_y(t_n,y_n)y_n'' = f_{tt}(t_n,y_n) + 2f_{ty}(t_n,y_n)y_n' + f_{yy}(t_n,y_n)(y_n')^2 +$
    $f_y(t_n,y_n)y_n''$. Using the theory of Taylor series with a
    remainder, we find the local formula error will be

    $\phi''''(\bar{t}_n)h^4/4!$ where $t_n < \bar{t}_n < t_{n+1}$ and ϕ is the exact
    solution of the I.V.P.

Section 8.4, Page 409

2a. The Runge-Kutta formula is
    $y_{n+1} = y_n + h(k_{n1} + 2k_{n2} + 2k_{n3} + k_{n4})/6$ where k_{n1}, k_{n2}
    etc. are given by Eqs.(3). Thus for
    $f(t,y) = 0.5 - t + 2y$, $(t_0,y_0) = (0,1)$ and $h = .1$ we have
    $k_{01} = f(0,1) = .5 + 2 = 2.5$
    $k_{02} = f(.05, 1.125) = .5 - .05 + 2.25 = 2.7$
    $k_{03} = f(.05, 1.135) = .5 - .05 + 2.27 = 2.72$
    $k_{04} = f(.1, 1.272) = .5 - .1 + 2.544 = 2.944$
    and hence
    $y(.1) \cong y_1 = 1 + .1(2.5 + 5.4 + 5.44 + 2.944)/6 = 1.2714$.
    To approximate $y(.2)$ we have
    $k_{11} = f(.1, 1.2714) = .5 - .1 + 2.5428 = 2.9428$
    $k_{12} = f(.15, 1.41854) = .5 - .15 + 2.83708 = 3.18708$
    $k_{13} = f(.15, 1.430754) = .5 - .15 + 2.861508 = 3.211508$
    $k_{14} = f(.2, 1.5925508) = .5 - .2 + 3.1851016 = 3.4851016$
    and thus
    $y(.2) \cong y_2 = 1.2714 + .1(k_{11} + 2k_{12} + 2k_{13} + k_{14})/6 = 1.59182$.

5a. We have, for $h = .1$, $t_0 = 1$ and $y_0 = 3$

$k_{01} = f(t_0, y_0) = \sqrt{t_0 + y_0} = 2$

$k_{02} = f(t_0+h/2,\ y_0 + hk_{01}/2) = \sqrt{t_0+.05+y_0+.1} = 2.03715$

$k_{03} = f(t_0+h/2,\ y_0+hk_{02}/2) = \sqrt{t_0+.05+y_0+.101858} = 2.03761$

$k_{04} = f(t_0+h,\ y_0+hk_{03}) = \sqrt{t_0+.1+y_0+2.03761} = 2.07455$

and thus

$y_1 = 3 + .1(2 + 4.07431 + 4.07522 + 2.07455)/6 = 3.20373$.

For Problems 9 through 12 use the BASIC program given in the text (or a comparable computer program) or a programmable calculator.

14a. The difference between formulas (i) and (ii) is $hy_n' +$
$h^2 y_n''/2 - h\{af(t_n, y_n) + bf[t_n + \alpha h,\ y_n + \beta h f(t_n, y_n)]\}$.
Since $y' = f(t_n, y_n)$ we have
$y_n'' = f_t(t_n, y_n) + f_y(t_n, y_n) f(t_n, y_n)$ and from Problem 15 of
Section 8.3 we find (note that $h = \alpha h$ and $k = \beta h f$)
$f[t_n + \alpha h,\ y_n + \beta h f(t_n, y_n)] = f(t_n, y_n) + \alpha h f_t(t_n, y_n) +$
$\beta h f(t_n, y_n) f_y(t_n, y_n) +$ terms involving h^2. Using these
expressions we find that if $a + b = 1$, $b\alpha = 1/2$ and $b\beta =$
$1/2$, then the difference reduces to h times the terms
involving h^2 and hence the difference is proportional to
h^3.

14b. There are only three equations involving four unknowns.
Choose b to be arbitrary ($b = \lambda$) and then solve for $a, b,$
and α in terms of λ.

15. We require that $At^2 + Bt + C$ equal $f(t)$ at $t = 0$,
$t = h/2$, and $t = h$. Hence $C = f(0)$,
$Ah^2/4 + Bh/2 + C = f(h/2)$, and $Ah^2 + Bh + C = f(h)$. The
solution of this system is
$A = 2[f(0) - 2f(h/2) + f(h)]/h^2$,
$B = [-3f(0) + 4f(h/2) - f(h)]/h$, and $C = f(0)$. Hence
$$\int_0^h f(t)\,dt \cong \int_0^h (At^2 + Bt + C)\,dt = Ah^3/3 + Bh^2/2 + Ch$$
$$= (h/6)[4f(0) - 8f(h/2) + 4f(h) - 9f(0) +$$
$$12f(h/2) - 3f(h) + 6f(0)]$$
$$= h[f(0) + 4f(h/2) + f(h)]/6.$$

Section 8.5, Page 415

2a. If $0 \leq t \leq 1$ then we know $0 \leq t^2 \leq 1$ and hence
     $e^y \leq t^2 + e^y \leq 1 + e^y$. Since each of these terms
     represents a slope, we may conclude that the solution of
     Eq.(i) is bounded above by the solution of Eq.(iii) and
     is bounded below by the solution of Eq.(iv).

2b. $\phi_1(t)$ and $\phi_2(t)$ can each be found by separation of
     variables. For $\phi_1(t)$ we have $\dfrac{1}{1+e^y}dy = dt$, or
     $\dfrac{e^{-y}}{e^{-y}+1}dy = dt$. Integrating both sides yields
     $-\ln(e^{-y}+1) = t + c$. Solving for y we find
     $y = \ln[1/(c_1e^{-t}-1)]$. Setting $t = 0$ and $y = 0$, we obtain
     $c_1 = 2$ and thus $\phi_1(t) = \ln[e^t/(2-e^t)]$. As $t \to \ln2$, we
     see that $\phi_1(x) \to \infty$. A similar analysis shows that
     $\phi_2(t) = \ln[1/(c_2-t)]$, where $c_2 = 1$ when the I.C. are
     used. Thus $\phi_2(t) \to \infty$ as $x \to 1$ and thus we conclude
     that $\phi(t) \to \infty$ for some t such that $\ln2 \leq t \leq 1$.

2c. Utilize the solutions found in Part b.

3a. The general solution of the D.E. is $y(t) = t + ce^{\lambda t}$,
     where $y(0) = 0 \to c = 0$ and thus $y(t) = t$, which is
     independent of λ.

3c. Your result in Part b will depend upon the particular
     computer hardware and software that you use. If there is
     sufficient accuracy, you will obtain the solution $y = t$
     for t on $0 \leq t \leq 1$ for each value of λ that is given,
     since there is no discretization error. If there is not
     sufficient accuracy, then round-off error will affect
     your calculations. For the larger values of λ, the
     numerical solution will quickly diverge from the exact
     solution, $y = t$, to the general solution $y = t + ce^{\lambda t}$,
     where the value of c depends upon the round-off error.
     If the latter case does not occur, you may simulate it by
     computing the numerical solution to the I.V.P.
     $y' - \lambda y = 1 - \lambda t$, $y(.1) = .10000001$. Here we have
     assumed that the numerical solution is exact up to the
     point $t = .09$ [i.e. $y(.09) = .09$] and that $t = .1$ round-
     off error has occurred as indicated by the slight error
     in the I.C. It has also been found that a larger step
     size ($h = .05$ or $h = .1$) may also lead to round-off
     error.

Section 8.6, Page 421

2.   The predictor formula is
$$y_{n+1} = y_n + (h/24)(55y'_n - 59y'_{n-1} + 37y'_{n-2} - 9y'_{n-3})$$
and the corrector formula is
$$y_{n+1} = y_n + (h/24)(9y'_{n+1} + 19y'_n - 5y'_{n-1} + y'_{n-2}),\text{ where}$$
$y'_n = .5 - t_n + 2y_n$. From Section 8.4, Problem 2 we have
$y_1 = 1.2714$, $y_2 = 1.59182$ and $y_3 = 1.97211$, so $y'_0 = 2.5$,
$y'_1 = 2.9428$, $y'_2 = 3.48364$ and $y'_3 = 4.14422$. Thus the
predicted value of y_4 is
$$y_4 = 1.97211 + (.1/24)[55(4.14422) - 59(3.48364) +$$
$37(2.9428) - 9(2.5)] = 2.425364$. This gives $y'_4 = 4.95073$
and thus the corrected value of y_4 is
$$y_4 = 1.97211 + (.1/24)[9(4.95073) + 19(4.14422) -$$
$5(3.48364) + 2.9428] = 2.42553$. This gives a corrected
value for $y'_4 = 4.95106$. The predicted value for y_5 is
now given as
$$y_5 = 2.42553 + (.1/24)[55(4.95106) - 59(4.14422) +$$
$37(3.48364) - 9(2.9428)] = 2.968067$. This gives
$y'_5 = 5.593613$ and thus
$$y_5 = 2.42553 + (.1/24)[9(5.93613) + 19(4.95106) -$$
$5(4.14422) + 3.48364] = 2.96827$ is the corrected value.

For Problems 9 through 12, the values for y_1, y_2 and y_3 were
obtained in Problems 9 through 12 of Section 8.4. As
before, it would probably be best to write a computer
program to solve these problems.

13.   With $y' = at + b$ we require that $y'_{n-1} = y'(t_{n-1}) =$
$at_{n-1} + b = a(t_n - h) + b$ and $y'_n = y'(t_n) = at_n + b$.
Solving the linear system for a and b yields
$a = (y'_n - y'_{n-1})/h$, $b = [t_n y'_{n-1} - (t_n - h)y']/h$. Next
$$\phi(t_n+1) - \phi(t_{n-1}) = \int_{t_{n-1}}^{t_{n+1}} \phi'(t)\,dt \cong \int_{t_n-h}^{t_n+h} (at+b)\,dt = 2aht_n + 2bh$$
$$= 2(y'_n - y'_{n-1})t_n + 2[t_n y'_{n-1} - (t_n-h)y'_n] = 2hy'_n.$$
Setting $\phi(t_n) = y_n$, it follows that $y_{n+1} - y_{n-1} = 2hy'_n$.

14.   We have, using Taylor series:
$$\phi(t_n+h) = \phi(t_n) + h\phi'(t_n) + \phi''(t_n)h^2/2 + \phi'''(\alpha_n)h^3/6$$
and
$$\phi(t_n-h) = \phi(t_n) - h\phi'(t_n) + \phi''(t_n)h^2/2 - \phi'''(\beta_n)h^3/6,\text{ where}$$
$t_n < \alpha_n < t_n + h$ and $t_n - h < \beta_n < t_n$. Subtracting the
second equation from the first, we obtain
$$\phi(t_n+h) - \phi(t_n-h) = 2h\phi'(t_n) + [\phi'''(\alpha_n) + \phi'''(\beta_n)]h^3/6.$$

From Problem 13 we have $y_{n+1} = y_{n-1} + 2hy'_n$ and thus $e_{n+1} = \phi(t_n+h) - y_{n+1} = [\phi'''(\alpha_n) + \phi'''(\beta_n)]h^3/6$, since at each step it is assumed that $\phi(t_n-h) = y_{n-1}$ and $\phi'(t_n) = y'_n$. Hence the local discretization error is proportional to h^3 plus possibly higher order terms.

15b. The predictor formula is
$y_{n+1} = y_{n-3} + (4h/3)[2y'_n - y'_{n-1} + 2y'_{n-2}]$ and the corrector formula is $y_{n+1} = y_{n-1} + (h/3)[y'_{n-1} + 4y'_n + y'_{n+1}]$, where $y'_n = 1/2 - t_n + 2y_n$. For the predictor formula we may use the values of y'_n, y'_{n-1} and y'_{n-2} calculated in Problem 2 and thus
$y_{4p} = 1 + (.4/3)[2(4.14422) - 3.48364 + 2(2.9428)] = 2.42539$.
Using this value for y_4 we find that $y'_4 = 4.95077$ and thus
$y_{4c} = 1.59182 + (.1/3)[3.48364 + 4(4.14422) + 4.95078]$
$\quad\quad = 2.42553$.

16b. Substituting $y = c\exp(At)$ in Eq.(5) we find that
$y_{n+1} = y_n + (h/24)(9Ay_{n+1} + 19Ay_n - 5Ay_{n-1} + Ay_{n-2})$ or
$(1-9Ah/24)y_{n+1} - (1+19Ah/24)y_n + 5Ahy_{n-1}/24 - Ahy_{n-2}/24 = 0$,
where we may now set $\alpha = Ah/24$.

16c. Substituting $y_n = \lambda^n$ in Eq.(i) we have
$(1-9\alpha)\lambda^{n+1} - (1+19\alpha)\lambda^n + 5\alpha\lambda^{n-1} - \alpha\lambda^{n-2} =$
$[(1-9\alpha)\lambda^3 - (1+19\alpha)\lambda^2 + 5\alpha\lambda - \alpha]\lambda^{n-2} = 0$.

16d. If we let $\alpha \to 0$ in Eq.(ii) we obtain $\lambda^3 - \lambda^2 = 0$ and hence the roots are $\lambda_1 = 1$, $\lambda_2 = 0$, $\lambda_3 = 0$.

16e. Substituting $\lambda_1 = 1 + \lambda_{11}h$, $\lambda_1^2 = 1 + 2\lambda_{11}h + \lambda_{11}^2h^2$, and $\lambda_1^3 = 1 + 3\lambda_{11}h + 3\lambda_{11}^2h^2 + \lambda_{11}^3h^3$ in Eq.(ii) and neglecting terms proportional to h^2 and h^3, we have
$(1-9\alpha)(1+3\lambda_{11}h) - (1+19\alpha)(1+2\lambda_{11}h) + 5\alpha(1+\lambda_{11}h) - \alpha =$
$1 + 3\lambda_{11}h - 9\alpha - 1 - 2\lambda_{11}h - 19\alpha + 5\alpha - \alpha = 0$ (since $\alpha = Ah/24$ we have neglected products of α and h).
Simplifying this last equation we obtain $\lambda_{11}h = 24\alpha = Ah$ and thus $\lambda_{11} = A$. Hence $\lambda_1 \cong 1 + Ah$ which means $y_1 = \lambda_1^n \cong (1 + Ah)^n$. Now suppose we choose a point $t = t_n = nh$ and let $n \to \infty$ with $h \to 0$ so that $nh = t_n$ is fixed. Then $y_1 \cong (1+hA)^n = (1+At_n/n)^n \to \exp(At_n)$ as $h\to 0$.

Section 8.7, Page 425

2a. The Euler formula is $x_{n+1} = x_n + hx_n'$, $y_{n+1} = y_n + hy_n'$
     where $x_n' = 2x_n + t_ny_n$ and $y_n' = x_ny_n$, $n = 0,1,2...$ with
     $x_0 = 1, y_0 = 1$. Thus $x_1 = 1 + .1(2+0) = 1.2$,
     $y_1 = 1 + .1[(1)(1)] = 1.1$ and
     $x_2 = 1.2 + .1[2.4 + (.1)(1.1)] = 1.451$,
     $y_2 = 1.1 + .1[(1.2)(1.1)] = 1.232$.

2b. We have $f(t,x,y) = 2x + ty$, $g(t,x,y) = xy$, $t_0 = 0$, $x_0 = 1$
     and $y_0 = 1$. Thus Eqs.(6) and (7) give
     $k_0 = 2(1) + (0)(1) = 2$, $\ell_{01} = (1)(1) = 1$
     $k_{02} = 2(1 + .2) + (.1)(1 + .1) = 2.51$,
     $\ell_{02} = (1.2)(1.1) = 1.32$
     $k_{03} = 2(1 + .251) + (.1)(1 + .132) = 2.6152$
     $\ell_{03} = (1.251)(1.132) = 1.41613$
     $k_{04} = 2(1 + .52304) + (.2)(1 + .28323) = 3.30273$
     $\ell_{04} = (1.52304)(1.28323) = 1.95440$.
     Eqs.(5) now give
     $x_1 = 1+(.2/6)[2+2(2.51)+2(2.6152)+3.30273] = 1.51844$
     $y_1 = 1+(.2/6)[1+2(1.32)+2(1.41613)+1.95440] = 1.28089$.

7.   Write a computer program to do this problem as there are
     twenty steps or more for $h \le .05$.

8.   If we let $y = x'$, then $y' = x''$ and thus we obtain the
     system $x' = y$ and $y' = t-3x-t^2y$, with $x(0) = 1$ and
     $y(0) = x'(0) = 2$. Thus $f(t,x,y) = y$,
     $g(t,x,y) = t - 3x - t^2y$, $t_0 = 0$, $x_0 = 1$ and $y_0 = 2$. If a
     program has been written for an earlier problem, then its
     best to use that. Otherwise, the first two steps are as
     follows:
     $k_{01} = 2$, $\ell_{01} = -3$, $k_{02} = 2+(-.15) = 1.85$,
     $\ell_{02} = .05 - 3(1+.1) - (.05)^2(2-.15) = -3.25463$,
     $k_{03} = 2 + (-.16273) = 1.83727$,
     $\ell_{03} = .05 - 3(1+.0925) - (.05)^2(2-.16273) = -3.23209$,
     $k_{04} = 2 + (-.32321) = 1.67679$,
     $\ell_{04} = .1 - 3(1.18373) - (.1)^2(1.67679) = -3.46796$,
     $x_1 = 1+(.1/6)[2 + 2(1.85)+2(1.83727)+(1.67679)]=1.18419$,
     $y_1 = 2+(.1/6)[-3-2(3.25463)-2(3.23209)-3.46796]=1.67598$.
     The last two values are approximations to $x(.1)$ and
     $y(.1) = x'(.1)$. In a similar fashion we have

$k_{11} = y_1$, $\ell_{11} = .1-3x_1-(.1)^2y_1 = -3.46933$,

$k_{12} = y_1 + .05\ell_{11} = 1.50251$,

$\ell_{12} = .15-3(x_1+.05k_{11})-(.15)^2(1.50251) = -3.68777$,

$k_{13} = y_1 + .05\ell_{12} = 1.49159$,

$\ell_{13} = .15-3(x_1+.05k_{12})-(.15)^2(1.49159) = -3.66151$,

$k_{14} = y_1+.1\ell_{13} = 1.30983$, and

$\ell_{14} = .2-3(x_1+.1k_{13})-(.2)^2(1.30983) = -3.85244$, so

$x(.2) \cong x_2 = x_1+(.1/6)(k_{11}+2k_{12}+2k_{13}+k_{14}) = 1.33376$

$y(.2) \cong y_2 = y_1+(.1/6)(\ell_{11}+2\ell_{12}+2\ell_{13}+\ell_{14}) = 1.30897$. Three more steps must be taken in order to approximate $x(.5)$ and $y(.5) = x'(.5)$. The intermediate steps yield $x(.3) \cong 1.44489$, $y(.3) \cong .9093062$ and $x(.4) \cong 1.51499$, $y(.4) \cong .4908795$.

CHAPTER 9

Section 9.1, Page 437

For Problems 1 through 16, once the eigenvalues have been
found, Table 9.1.1 will for the most part quickly yield the
type of critical point and the stability. In all cases it
can be easily verified that $\underset{\sim}{A}$ is nonsingular.

1a. The eigenvalues are found from the equation $\det(\underset{\sim}{A}-r\underset{\sim}{I})=0$.

Substituting the values for $\underset{\sim}{A}$ we have $\begin{vmatrix} 3-r & -2 \\ 2 & -2-r \end{vmatrix} =$

$r^2 - r + 2 = 0$ and thus the eigenvalues are $r_1 = -1$ and

$r_2 = 2$. For $r_1 = -1$, we have $\begin{pmatrix} 4 & -2 \\ 2 & -1 \end{pmatrix}\begin{pmatrix} \xi_1 \\ \xi_2 \end{pmatrix} = \begin{pmatrix} 0 \\ 0 \end{pmatrix}$ and thus

$\underset{\sim}{\xi}^{(1)} = \begin{pmatrix} 1 \\ 2 \end{pmatrix}$ and for r_2 we have $\begin{pmatrix} 1 & -2 \\ 2 & -4 \end{pmatrix}\begin{pmatrix} \xi_1 \\ \xi_2 \end{pmatrix} = \begin{pmatrix} 0 \\ 0 \end{pmatrix}$ and thus

$\underset{\sim}{\xi}^{(2)} = \begin{pmatrix} 2 \\ 1 \end{pmatrix}$.

1b. Since the eigenvalues differ in sign, the critical point
is a saddle point and is unstable.

1d.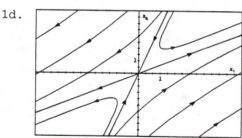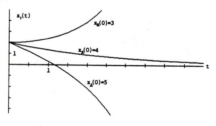

4a. Again the eigenvalues are given by $\begin{vmatrix} 1-r & -4 \\ 4 & -7-r \end{vmatrix} =$

$r^2 + 6r + 9 = 0$ and thus $r_1 = r_2 = -3$. The eigenvectors

are solutions of $\begin{pmatrix} 4 & -4 \\ 4 & -4 \end{pmatrix}\begin{pmatrix} \xi_1 \\ \xi_2 \end{pmatrix} = \begin{pmatrix} 0 \\ 0 \end{pmatrix}$ and hence there is

just one eigenvector $\underset{\sim}{\xi} = \begin{pmatrix} 1 \\ 1 \end{pmatrix}$.

4b. Since the eigenvalues are negative, $(0,0)$ is an improper
node which is asymptotically stable. If we had found

that there were two independent eigenvectors then $(0,0)$
would have been a proper node, as indicated in Case 3a.

4d.

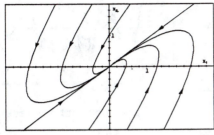

7a. In this case $\det(\underset{\sim}{A} - r\underset{\sim}{I}) = r^2 - 2r + 5$ and thus the
eigenvalues are $r_{1,2} = 1 \pm 2i$. For $r_1 = 1 + 2i$ we have
$$\begin{pmatrix} 2-2i & -2 \\ 4 & -2-2i \end{pmatrix} \begin{pmatrix} \xi_1 \\ \xi_2 \end{pmatrix} = \begin{pmatrix} 2-2i & -2 \\ 8-8i & -8 \end{pmatrix} \begin{pmatrix} \xi_1 \\ \xi_2 \end{pmatrix} = \begin{pmatrix} 0 \\ 0 \end{pmatrix} \text{ and thus}$$
$\underset{\sim}{\xi}^{(1)} = \begin{pmatrix} 1 \\ 1-i \end{pmatrix}$. Similarly for $r_2 = 1-2i$ we have
$$\begin{pmatrix} 2+2i & -2 \\ 4 & -2+2i \end{pmatrix} \begin{pmatrix} \xi_1 \\ \xi_2 \end{pmatrix} = \begin{pmatrix} 0 \\ 0 \end{pmatrix} \text{ and hence } \underset{\sim}{\xi}^{(2)} = \begin{pmatrix} 1 \\ 1+i \end{pmatrix}.$$

7b. Since the eigenvalues are complex with positive real
part, we conclude that the critical point is a spiral
point and is unstable.

7d.

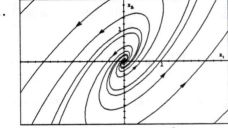

10a. Again, $\det(\underset{\sim}{A} - r\underset{\sim}{I}) = r^2 + 9$ and thus we have $r_{1,2} = \pm 3i$.
For $r_1 = 3i$ we have $\begin{pmatrix} 1-3i & 2 \\ -5 & -1-3i \end{pmatrix} \begin{pmatrix} \xi_1 \\ \xi_2 \end{pmatrix} = \begin{pmatrix} 0 \\ 0 \end{pmatrix}$ and thus
$\underset{\sim}{\xi}^{(1)} = \begin{pmatrix} 2 \\ 1-3i \end{pmatrix}$. Likewise for $r_2 = -3i$,
$$\begin{pmatrix} 1+3i & 2 \\ -5 & -1+3i \end{pmatrix} \begin{pmatrix} \xi_1 \\ \xi_2 \end{pmatrix} = \begin{pmatrix} 0 \\ 0 \end{pmatrix} \text{ so that } \underset{\sim}{\xi}^{(2)} = \begin{pmatrix} 2 \\ 1+3i \end{pmatrix}.$$

10b. Since the eigenvalues are pure imaginary the critical
point is a center, which is stable.

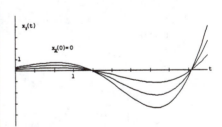

10d.

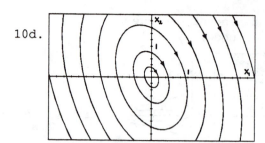 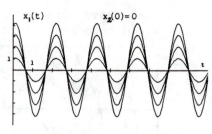

13. If we let $\underset{\sim}{x} = \underset{\sim}{x}^0 + \underset{\sim}{u}$ then $\underset{\sim}{x}' = \underset{\sim}{u}'$ and thus the system

becomes $\underset{\sim}{u}' = \begin{pmatrix} 1 & 1 \\ 1 & -1 \end{pmatrix} \underset{\sim}{x}^0 + \begin{pmatrix} 1 & 1 \\ 1 & -1 \end{pmatrix} \underset{\sim}{u} - \begin{pmatrix} 2 \\ 0 \end{pmatrix}$ which will be in

the form of Eq. (2) if $\begin{pmatrix} 1 & 1 \\ 1 & -1 \end{pmatrix} \underset{\sim}{x}^0 = \begin{pmatrix} 2 \\ 0 \end{pmatrix}$. Using row

operations, this last set of equations is equivalent to
$\begin{pmatrix} 1 & 1 \\ 0 & -2 \end{pmatrix} \underset{\sim}{x}^0 = \begin{pmatrix} 2 \\ -2 \end{pmatrix}$ and thus $x_1^0 = 1$ and $x_2^0 = 1$. Since

$\underset{\sim}{u}' = \begin{pmatrix} 1 & 1 \\ 1 & -1 \end{pmatrix} \underset{\sim}{u}$ has $(0,0)$ as the critical point, we

conclude that $(1,1)$ is the critical point of the original
system. As in the earlier problems, the eigenvalues are

given by $\begin{vmatrix} 1-r & 1 \\ 1 & -1-r \end{vmatrix} = r^2 - 2 = 0$ and thus $r_{1,2} = \pm\sqrt{2}$.

Hence the critical point $(1,1)$ is an unstable saddle
point.

17. The equivalent system is $dx/dt = y, dy/dt = -(k/m)x-(c/m)y$
which is written in the form of Eq. (2) as
$\frac{d}{dt}\begin{pmatrix} x \\ y \end{pmatrix} = \begin{pmatrix} 0 & 1 \\ -k/m & -c/m \end{pmatrix}\begin{pmatrix} x \\ y \end{pmatrix}$. The point $(0,0)$ is clearly a

critical point, and since $\underset{\sim}{A}$ is nonsingular, it is the

only one. The characteristic equation is $r^2+(c/m)r+k/m=0$
so $r_1, r_2 = [-c \pm (c^2 - 4km)^{1/2}]/2m$. In the underdamped
case $c^2 - 4km < 0$, the characteristic roots are complex
with negative real parts and thus the critical point
$(0,0)$ is an asymptotically stable spiral point. In the
overdamped case $c^2 - 4km > 0$, the characteristic roots
are real, unequal, and negative and hence the critical
point $(0,0)$ is asymptotically stable improper node. In
the critically damped case $c^2 - 4km = 0$, the
characteristic roots are equal and negative. As
indicated in the solution to Problem 4, to determine

whether this is an improper or proper node we must
determine whether there are one or two linearly
independent eigenvectors. The eigenvectors satisfy the

equations $\begin{pmatrix} c/2m & 1 \\ -k/m & -c/2m \end{pmatrix} \begin{pmatrix} \xi_1 \\ \xi_2 \end{pmatrix} = \begin{pmatrix} 0 \\ 0 \end{pmatrix}$, which have just one

solution if $c^2 - 4km = 0$. Thus the critical point $(0,0)$
is an asymptotically stable improper node.

18a. If $\underset{\sim}{A}$ has one zero eigenvalue then for $r = 0$ we have

$\det(\underset{\sim}{A} - r\underset{\sim}{I}) = \det\underset{\sim}{A} = 0$. Hence $\underset{\sim}{A}$ is singular which means

$\underset{\sim}{A}\underset{\sim}{x} = \underset{\sim}{0}$ has infinitely many solutions and consequently

there are infinitely many critical points.

18b. From Chapter 7, the solution is $\underset{\sim}{x}(t) = c_1\underset{\sim}{\xi}^{(1)} + c_2\underset{\sim}{\xi}^{(2)}e^{r_2 t}$,

which can be written in scalar form as
$x_1 = c_1\xi_1^{(1)} + c_2\xi_1^{(2)}e^{r_2 t}$ and $x_2 = c_1\xi_2^{(1)} + c_2\xi_2^{(2)}e^{r_2 t}$.
Assuming $\xi_1^{(2)} \neq 0$, the first equation can be solved for
$c_2 e^{r_2 t}$, which is then substituted into the second
equation to yield $x_2 = c_1\xi_2^{(1)} + [\xi_2^{(2)}/\xi_1^{(2)}][x_1 - c_1\xi_1^{(1)}]$.
These are straight lines parallel to the vector $\underset{\sim}{\xi}^{(2)}$.

Note that the family of lines is independent of c_2. If
$\xi_1^{(2)} = 0$, then the lines are vertical. If $r_2 > 0$, the
direction of motion will be in the same direction as
indicated for $\underset{\sim}{\xi}^{(2)}$. If $r_2 < 0$, then it will be in the

opposite direction.

19b. Eq.(i) can be written in scalar form as $dx/dt = ax + by$
and $dy/dt = cx + dy$, which then yields Eq.(iii).
Ignoring the middle quotient in Eq.(iii), we can rewrite
that equation as $(cx + dy)dx - (ax + by)dy = 0$, which is
exact since $d = -a$.

19c. Integrating $\phi_x = cx + dy$ we obtain $\phi = cx^2/2 + dxy + g(y)$
and thus $dx + dg/dy = -az - by$, or $dg/dy = -by$. Hence
$cx^2/2 + dxy - by^2/2 = k/2$ is the solution to Eq.(iii).
The quadratic equation $Ax^2 + Bxy + Cy^2 = 0$ is an ellipse
provided $B^2 - 4AC < 0$. Hence for our problem if
$4d^2 + 4bc < 0$ then Eq.(iv) is an ellipse. Using $a + d = 0$
we have $d^2 = -ad$ and hence $-ad + bc < 0$ or $ad - bc > 0$,
which is true by Eqs.(ii). Thus Eq.(iv) is an ellipse
under the conditons of Eqs.(ii).

20. The given system can be written as $\frac{d}{dt}\begin{pmatrix} x \\ y \end{pmatrix} = \begin{pmatrix} a & b \\ c & d \end{pmatrix}\begin{pmatrix} x \\ y \end{pmatrix}$.

Thus the eigenvalues are given by $r^2 - (a+d)r + ad-bc = 0$ and using the given definitions we rewrite this as $r^2 - pr + q = 0$ and thus

$r_{1,2} = (p \pm \sqrt{p^2-4q})/2 = (p \pm \sqrt{\Delta})/2$. The results are now obtained using Table 9.1.1.

Section 9.2, Page 446

1. Solutions of the D.E. for x are y are $x = Ae^{-t}$ and $y = Be^{-2t}$ respectively. $x(0) = 4$ and $y(0) = 2$ yield $A = 4$ and $B = 2$, so $x = 4e^{-t}$, and $y = 2e^{-2t}$. Solving the first equation for e^{-t} and then substituting into the second

yields $y = 2[x/4]^2 = x^2/8$, which is a parabola. From the original D.E., or from the parametric solutions, we find that $0 < x \leq 4$ and $0 < y \leq 2$ for $t \geq 0$ and thus only the portion of the parabola shown is the trajectory, with the direction of motion indicated.

3. Utilizing the approach indicated in Eq.(11), we have $dy/dx = -x/y$, which separates into $xdx + ydy = 0$. Integration then yields the circle $x^2 + y^2 = c^2$, where $c^2 = 16$ for both sets of I.C. The direction of motion can be found from the original D.E. and is counterclockwise for both I.C. To obtain the parametric equations, we write the system in the form

$\frac{d}{dt}\begin{pmatrix} x \\ y \end{pmatrix} = \begin{pmatrix} 0 & -1 \\ 1 & 0 \end{pmatrix}\begin{pmatrix} x \\ y \end{pmatrix}$, which has the characteristic

equation $\begin{vmatrix} -r & -1 \\ 1 & -r \end{vmatrix} = r^2 + 1 = 0$, or $r = \pm i$. Following

the procedures of Section 7.6, we find that one solution

of the above system is $\begin{pmatrix} 1 \\ -i \end{pmatrix}e^{it} = \begin{pmatrix} \cos t + i\sin t \\ \sin t - i\cos t \end{pmatrix}$ and thus

two real solutions are $\underset{\sim}{u}(t) = \begin{pmatrix} \cos t \\ \sin t \end{pmatrix}$ and $\underset{\sim}{v}(t) = \begin{pmatrix} \sin t \\ -\cos t \end{pmatrix}$.

The general solution of the system is then

$\begin{pmatrix} x \\ y \end{pmatrix} = c_1 \underset{\sim}{u}(t) + c_2 \underset{\sim}{v}(t)$ and hence the first I.C. yields $c_1 = 4$, $c_2 = 0$, or $x = 4\cos t$, $y = 4\sin t$. The second I.C. yields $c_1 = 0$, $c_2 = -4$, or $x = -4\sin t$, $y = 4\cos t$. Note that both these parametric representations satsify the form of the trajectories found in the first part of this problem.

6. The critical points are given by the solutions of $x(1-x-y) = 0$ and $y(1/2 - y/4 - 3x/4) = 0$. The solutions corresponding to either $x = 0$ or $y = 0$ are seen to be $x = 0$, $y = 0$; $x = 0$, $y = 2$; $x = 1$, $y = 0$. In addition, there is a solution corresponding to the intersection of the lines $1 - x - y = 0$ and $1/2 - y/4 - 3x/4 = 0$ which is the point $x = 1/2$, $y = 1/2$. Thus the critical points are $(0,0)$, $(0,2)$, $(1,0)$, and $(1/2,1/2)$.

9. We know that $\phi'(t) = F[\phi(t), \psi(t)]$ and $\psi'(t) = G[\phi(t), \psi(t)]$ for $\alpha < t < \beta$. By direct substitution we have $\Phi'(t) = \phi'(t-s) = F[\phi(t-s), \psi(t-s)] = F[\Phi(t), \psi(t)]$ and $\psi'(t) = \psi'(t-s) = G[\phi(t-s), \psi(t-s)] = G[\Phi(t), \psi(t)]$ for $\alpha < t-s < \beta$ or $\alpha+s < t < \beta+s$.

11. If $F(x_0,y_0,t) = G(x_0,y_0,t) = 0$ for all t then the conclusion is obvious. In a region where $F(x,y,t) \neq 0$ we may write

$$\frac{dy}{dx} = \frac{dy/dt}{dx/dt} = \frac{G(x,y,t)}{F(x,y,t)} = \frac{G(x,y,t)/[F^2(x,y,t) + G^2(x,y,t)]^{1/2}}{F(x,y,t)/[F^2(x,y,t) + G^2(x,y,t)]^{1/2}}$$
$$= \frac{A(x,y)}{B(x,y)}.$$

This is an autonomous D.E. which yields a one-parameter family of solutions independent of s. This family is the set of trajectories of the original system. A similar argument holds near points where $F(x,y,t) = 0$ but $G(x,y,t) \neq 0$ if we consider $dx/dy = F(x,y,t)/G(x,y,t)$.

12. Letting $x = \theta$ and $y = dx/dt$, then the D.E. can be written as the system $dx/dt = y$, $dy/dt = -(g/l)\sin x$. Using the approach of Eq.(11), we find that this system has $dy/dx = (-g/l)\sin(x/y)$, which separates into $ydy = (-g/l)\sin x dx$. Integrating both sides yields $y^2/2 = (g/l)\cos x + c_1$. Setting $c_1 = c - g/l$ then yields the desired form.

15. Suppose that $t_1 > t_0$. Let $s = t_1 - t_0$. Since the system
    is autonomous, the result of Problem 9, with s replaced
    by −s shows that $x = \phi_1(t+s)$ and $y = \psi_1(t+s)$ generates
    the same trajectory (C_1) as $x = \phi_1(t)$ and $y = \psi_1(t)$. But
    at $t = t_0$ we have $x = \phi_1(t_0+s) = \phi_1(t_1) = x_0$ and
    $y = \psi_1(t_0+s) = \psi_1(t_1) = y_0$. Thus the solution
    $x = \phi_1(t+s)$, $y = \psi_1(t+s)$ satisfies <u>exactly</u> the same
    initial conditions as the solution $x = \phi_0(t)$, $y = \psi_0(t)$
    which generates the trajectory C_0. Hence C_0 and C_1 are
    the same.

16. From the existence and uniqueness theorem we know that if
    the two solutions $x = \phi(t)$, $y = \psi(t)$ and $x = x_0$, $y = y_0$
    satisfy $\phi(a) = x_0$, $\psi(a) = y_0$ and $x = x_0$, $y = y_0$ at $t = a$,
    then these solutions are identical. Hence $\phi(t) = x_0$ and
    $\psi(t) = y_0$ for all t contradicting the fact that the
    trajectory generated by $[\phi(t), \psi(t)]$ started at a
    noncritical point.

17. By direct substitution $\Phi'(t) = \phi'(t+T) =$
    $F[\phi(t+T), \psi(t+T)] = F[\Phi(t), \Psi(t)]$ and $\Psi'(t) = \psi'(t+T) =$
    $G[\phi(t+T), \psi(t+T)]$, $G[\Phi(t), \psi(t)]$. Furthermore
    $\Phi(t_0) = x_0$ and $\Psi(t_0) = y_0$. Thus by the existence and
    uniqueness theorem $\Phi(t) = \phi(t)$ and $\Psi(t) = \psi(t)$ for all
    t.

Section 9.3, Page 456

In Problems 1 through 10, write the system in the form of
Eq.(4). Then if $\underset{\sim}{g}(\underset{\sim}{0}) = \underset{\sim}{0}$ we may conclude that (0,0) is a

critical point. In addition, if $\underset{\sim}{g}$ satisfies Eq.(5) or

Eq.(6), then the system is almost linear. In this case the
linear system, Eq.(1), will determine, in most cases, the
type and stability of the critical point (0,0) of the almost
linear system. These results are summarized in Table 9.3.1.

1. In this case the system can be written as
    $$\frac{d}{dt}\begin{pmatrix} x \\ y \end{pmatrix} = \begin{pmatrix} 1 & -1 \\ 3 & -2 \end{pmatrix}\begin{pmatrix} x \\ y \end{pmatrix} + \begin{pmatrix} xy \\ -xy \end{pmatrix} \text{ and thus } \underset{\sim}{A} = \begin{pmatrix} 1 & -1 \\ 3 & -2 \end{pmatrix} \text{ and }$$
    $\underset{\sim}{g} = \begin{pmatrix} xy \\ -xy \end{pmatrix}$. Since $\underset{\sim}{g}(\underset{\sim}{0}) = \begin{pmatrix} 0 \\ 0 \end{pmatrix}$ we conclude that (0,0) is a
    critical point. Following the procedure of Example 1, we
    let $x = r\cos\theta$ and $y = r\sin\theta$ and thus

$g_1(x,y)/r = -g_2(x,y)/r = \dfrac{r^2\cos\theta\sin\theta}{r} \to 0$ as $r \to 0$ and

thus the system is almost linear. Since
$\det(A-rI) = r^2 + r + 1$, we find that the eigenvalues are
$r_{1,2} = -1/2 \pm \sqrt{3}\,i/2$ and thus $(0,0)$ is an asymptotically
stable spiral point.

2. We have $\dfrac{d}{dt}\begin{pmatrix} x \\ y \end{pmatrix} = \begin{pmatrix} 1 & 0 \\ 0 & 1 \end{pmatrix}\begin{pmatrix} x \\ y \end{pmatrix} + \begin{pmatrix} x^2 + y^2 \\ -xy \end{pmatrix}$ and since

$\underset{\sim}{g}(0) = \begin{pmatrix} 0 \\ 0 \end{pmatrix}$ we conclude that $(0,0)$ is a critical point.

Letting $x = r\cos\theta$ and $y = r\sin\theta$ we see that
$g_1(x,y)/r = (r^2\cos^2\theta + r^2\sin^2\theta)/r \to 0$ and that
$g_2(x,y)/r = r^2\cos\theta\sin\theta/r \to 0$ as $r \to 0$. Hence the system
is almost linear. The eigenvalues for the corresponding
linear system are found to be $r_1 = 1$ and $r_2 = 1$. For the
linear system we could determine whether $(0,0)$ was a
proper or improper node. However, for the almost linear
system the repeated eigenvalues will be perturbed and
thus we cannot determine whether $(0,0)$ is a proper node,
improper node or a spiral point (if the perturbed values
are complex). In any case, since r_1 and r_2 are positive,
the system will be unstable.

8. The given system can be written as
$\dfrac{d}{dt}\begin{pmatrix} x \\ y \end{pmatrix} = \begin{pmatrix} 0 & 1 \\ 0 & 1 \end{pmatrix}\begin{pmatrix} x \\ y \end{pmatrix} + \begin{pmatrix} 1 - e^{-x} \\ -\sin x \end{pmatrix}$. In this case, though,

$\underset{\sim}{A} = \begin{pmatrix} 0 & 1 \\ 0 & 1 \end{pmatrix}$ is a singular and $\underset{\sim}{g} = \begin{pmatrix} 1 - e^{-x} \\ -\sin x \end{pmatrix}$ does not

satisfy Eq.(6) and thus we may not proceed. However, if
we consider the Taylor series for e^{-x} and for $\sin x$, we
see that $1 - x - e^{-x} = \dfrac{-x^2}{2!} + \dfrac{x^3}{3!} + \dots$ and

$-\sin x + x = \dfrac{x^3}{3!} - \dfrac{x^5}{5!} + \dots$ and hence if we let $x = r\cos\theta$
we may conclude that $(1 - x - e^{-x})/r \to 0$ and
$(-\sin x + x)/r \to 0$ as $r \to 0$. Using this information we
now add and subtract x to the first equation to obtain
$\dfrac{dx}{dt} = x + y + 1 - x - e^{-x}$ and similarly for the second

equation to obtain $\dfrac{dy}{dt} = -x + y - \sin x + x$. This equivalent system can now be written as

$$\frac{d}{dt}\begin{pmatrix} x \\ y \end{pmatrix} = \begin{pmatrix} 1 & 1 \\ -1 & 1 \end{pmatrix}\begin{pmatrix} x \\ y \end{pmatrix} + \begin{pmatrix} 1 - x - e^{-x} \\ -\sin x + x \end{pmatrix}.$$ Clearly, (0,0) is

a critical point and the system is almost linear by the

above analysis. The linear system, with $A = \begin{pmatrix} 1 & 1 \\ -1 & 1 \end{pmatrix}$, has

the eigenvalues $r_{1,2} = 1 \pm i$ and hence the almost linear
system has an unstable spiral point at (0,0).

13. The critical points are the solutions of $x(1-x-y) = 0$ and
    $y(3-x-2y) = 0$. Solutions are $x = 0$, $y = 0$; $x = 0$,
    $3 - 2y = 0$ which give $y = 3/2$; $y = 0$ and $1 - x = 0$ which
    give $x = 1$; and $1 - x - y = 0$, $3 - x - 2y = 0$ which give
    $x = 1$, $y = 2$. Thus the critical points are (0,0),
    (0,3/2), (1,0) and (-1,2). For the critical point (0,0)
    the D.E. is already in the form of an almost linear
    system; and the corresponding linear system is $dx/dt = x$,
    $dy/dt = 3y$ which has the eigenvalues $r_1 = 1$ and $r_2 = 3$.
    Thus the critical point (0,0) is an unstable improper
    node. Each of the other three critical points is dealt
    with in the same manner; we consider only the critical
    point (-1,2). In order to translate this critical point
    to the origin we set $x(t) = -1 + u(t)$, $y(t) = 2 + v(t)$
    and substitute in the D.E. to obtain
    $du/dt = -1 + u - (-1+u)^2 - (-1+u)(2+v) = u + v - u^2 - uv$
    and

$dv/dt = 3(2+v) - (-1+u)(2+v) - 2(2+v)^2 = -2u - 4v - uv - 2v^2$.
    Writing this in the form of Eq.(4) we find that

$$A = \begin{pmatrix} 1 & 1 \\ -2 & -4 \end{pmatrix} \text{ and } g = -\begin{pmatrix} u^2 + uv \\ uv + v^2 \end{pmatrix} \text{ which is an almost linear}$$

system. The eigenvalues of the corresponding linear
    system are $r = (-3 \pm \sqrt{9 + 8})/2$ and hence the critical
    point (-1,2), of the original system, is an unstable
    saddle point.

16a. The system is $\dfrac{d}{dt}\begin{pmatrix} x \\ y \end{pmatrix} = \begin{pmatrix} 1 & 0 \\ 0 & -2 \end{pmatrix}\begin{pmatrix} x \\ y \end{pmatrix} + \begin{pmatrix} 0 \\ x^2 \end{pmatrix}$ and thus is
    almost linear using the procedures outlined in the
    earlier problems. The corresponding linear system has
    the eigenvalues $r_1 = 1$, $r_2 = -2$ and thus (0,0) is an
    unstable saddle point.

16b. The trajectories of the linear system are the solutions
    of $dx/dt = x$ and $dy/dt = -2y$ and thus $x(t) = c_1 e^t$ and
    $y(t) = c_2 e^{-2t}$. To sketch these, solve the first equation

for e^t and substitute into the second to obtain
$y = c_1^2 c_2/x^2$, $c_1 \neq 0$. Several
trajectories are shown in the
figure. Since $x(t) = c_1 e^t$,
we must pick $c_1 = 0$ for
$x \to 0$ and $t \to \infty$. Thus $x = 0$,
$y = c_2 e^{-2t}$ (the vertical axis)
is the only trajectory for
which $x \to 0$, $y \to 0$ as $t \to \infty$.

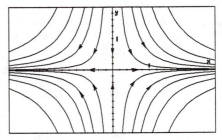

16c. For $x \neq 0$ we have $dy/dx = (dy/dt)/(dx/dt) = (-2y+x^3)/x$.
This is a linear equation, and the general solution is
$y = x^3/5 + k/x^2$, where k is an arbitrary constant. In
addition the system of equations has the solution $x = 0$,
$y = Be^{-2t}$. Any solution with its initial point on the
y-axis ($x=0$) is given by the latter solution. The
trajectories corresponding
to these solutions approach
the origin as $t \to \infty$. The
trajectory that passes through
the origin and divides the
family of curves is given by
$k = 0$, namely $y = x^3/5$. This
trajectory corresponds to the
trajectory $y = 0$ for the linear
problem. Several trajectories
are sketched in the figure.

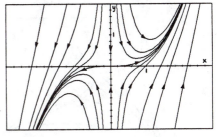

21. The equation of the trajectories was found in Problem 12
of Section 9.2.

22. Setting $c = 0$ in Eq.(18) of Section 9.2 and multiplying
by $m\ell^2$ we obtain $m\ell^2 d^2\theta/dt^2 + mg\ell\sin\theta = 0$. Considering
$d\theta/dt$ as a function of θ and using the chain rule we have
$$\frac{d}{dt}\left(\frac{d\theta}{dt}\right) = \frac{d}{d\theta}\left(\frac{d\theta}{dt}\right)\frac{d\theta}{dt} = \frac{1}{2}\frac{d}{d\theta}\left(\frac{d\theta}{dt}\right)^2 . \text{ Thus}$$
$(1/2)m\ell^2 d[(d\theta/dt)^2]/d\theta = -mg\ell\sin\theta$. Now integrate both
sides from α to θ where $d\theta/dt = 0$ at $\theta = \alpha$:
$(1/2)m\ell^2(d\theta/dt)^2 = mg\ell(\cos\theta - \cos\alpha)$. Thus
$(d\theta/dt)^2 = (2g/\ell)(\cos\theta - \cos\alpha)$. Since we are releasing
the pendulum with zero velocity from a positive angle α,
the angle θ will initially be decreasing so $d\theta/dt < 0$.
If we restrict attention to the range of θ from $\theta = \alpha$ to
$\theta = 0$, we can assert $d\theta/dt = -\sqrt{2g/\ell}\sqrt{\cos\theta - \cos\alpha}$.
Solving for dt gives $dt = -\sqrt{\ell/2g}\,d\theta/\sqrt{\cos\theta - \cos\alpha}$. Since

there is no damping, the pendulum will swing from its
initial angle α through 0 to $-\alpha$, then back through 0
again to the angle α in one period. It follows that
$\theta(T/4) = 0$. Integrating the last equation and noting
that as t goes from 0 to T/4, θ goes from α to 0 yields
$T/4 = -\sqrt{l/2g} \int_{\alpha}^{0} (1/\sqrt{\cos\theta - \cos\alpha})d\theta$. We obtain the
elliptic integral by making the change of variables given
in the text.

23b. Using the hint, the system can be written as
$$\frac{d}{dt}\begin{pmatrix} x \\ y \end{pmatrix} = \begin{pmatrix} 0 & 1 \\ -g'(0) & -c(0) \end{pmatrix}\begin{pmatrix} x \\ y \end{pmatrix} - \begin{pmatrix} 0 \\ -g''(\xi_1)x^2/2 - c'(\xi_2)xy \end{pmatrix}$$
where $0 < \xi_1, \xi_2 < x$, from which the results follow.

Section 9.4, Page 471

3d.

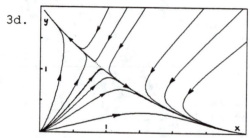

5a. The critical points are found by setting dx/dt = 0 and
dy/dt = 0 and thus we need to solve $x(1 - x - y) = 0$ and
$y(1.5 - y - x) = 0$. The first yields x = 0 or y = 1 - x
and the second yields y = 0 or y = 1.5 - x. Thus (0,0),
(0,3/2) and (1,0) are the only critical points since the
two straight lines do not intersect in the first quadrant
(or anywhere in this case). This is an example of one of
the cases shown in Figure 9.4.9 a or b.

5d.

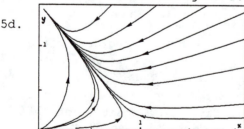

6a. The critical points are found by setting dx/dt = 0 and
dy/dt = 0 and thus we need to solve $x(1-x + y/2) = 0$ and
$y(5/2 - 3y/2 + x/4) = 0$. The first yields x = 0 or
y = 2x - 2 and the second yields y = 0 or y = x/6 + 5/3.

Thus we find the critical points $(0,0)$, $(1,0)$, $(0,5/3)$ and $(2,2)$. The last point is the intersection of the two straight lines, which will be used again in part c.

6b. For $(0,0)$ the linearized system is $x' = x$ and $y' = 5y/2$, which has the eigenvalues $r_1 = 1$ and $r_2 = 5/2$. Thus the origin is an unstable improper node. For $(2,2)$ we let $x = u + 2$ and $y = v + 2$ in the given system to find (since $x' = u'$ and $y' = v'$) that

$du/dt = (u+2)[1 - (u+2) + (v+2)/2] = (u+2)(-u+v/2)$ and

$dv/dt = (v+2)[5/2 - 3(v+2)/2 + (u+2)/4] = (v+2)(u/4 - 3v/2)$.

Hence the linearized equations are $\begin{pmatrix} u \\ v \end{pmatrix}' = \begin{pmatrix} -2 & 1 \\ 1/2 & -3 \end{pmatrix}\begin{pmatrix} u \\ v \end{pmatrix}$

which has the eigenvalues $r_{1,2} = (-5 \pm \sqrt{3})/2$. Since these are both negative we conclude that $(2,2)$ is an asymptotically stable improper node. In a similar fashion for $(1,0)$ we let $x = u + 1$ and $y = v$ to obtain the linearized system $\begin{pmatrix} u \\ v \end{pmatrix}' = \begin{pmatrix} -1 & 1/2 \\ 0 & 11/4 \end{pmatrix}\begin{pmatrix} u \\ v \end{pmatrix}$. This has

$r_1 = -1$ and $r_2 = 11/4$ as eigenvalues and thus $(1,0)$ is an unstable saddle point. Likewise, for $(0,5/3)$ we let

$x = u$, $y = v + 5/3$ to find $\begin{pmatrix} u \\ v \end{pmatrix}' = \begin{pmatrix} 11/6 & 0 \\ 5/12 & -5/2 \end{pmatrix}\begin{pmatrix} u \\ v \end{pmatrix}$ as the

corresponding linear system. Thus $r_1 = 11/6$ and $r_2 = -5/2$ and thus $(0,5/3)$ is an unstable saddle point.

6c. To sketch the required trajectories, we must find the eigenvectors for each of the linearized systems and then analyze the behavior of the linear solution near the critical point. Using this approach we find that the

solution near $(0,0)$ has the form $\begin{pmatrix} x \\ y \end{pmatrix} = c_1 \begin{pmatrix} 1 \\ 0 \end{pmatrix} e^t +$

$c_2 \begin{pmatrix} 0 \\ 1 \end{pmatrix} e^{5t/2}$ and thus the origin is approached only for

large negative values of t. In this case e^t dominates $e^{5t/2}$ and hence in the neighborhood of the origin all trajectories are tangent to the x-axis except for one pair ($c_1 = 0$) that lies along the y-axis. For $(2,2)$ we find the eigenvector corresponding to $r = (-5 + \sqrt{3})/2 = -1.63$ is given by $(1-\sqrt{3})\xi_1/2 + \xi_2 = 0$

and thus $\begin{pmatrix} 1 \\ (\sqrt{3}-1)/2 \end{pmatrix} = \begin{pmatrix} 1 \\ .37 \end{pmatrix}$ is one eigenvector. For

$r = (-5 -\sqrt{3})/2 = -3.37$ we have $(1 +\sqrt{3})\xi_1/2 + \xi_2 = 0$ and
thus $\begin{pmatrix} 1 \\ -(\sqrt{3}+1)/2 \end{pmatrix} = \begin{pmatrix} 1 \\ -1.37 \end{pmatrix}$ is the second eigenvector.

Hence the linearized solution is
$$\begin{pmatrix} u \\ v \end{pmatrix} = c_1 \begin{pmatrix} 1 \\ .37 \end{pmatrix} e^{-1.63t} + c_2 \begin{pmatrix} 1 \\ -1.37 \end{pmatrix} e^{-3.37t}. \quad \text{For large}$$

positive values of t the first term is the dominant one
and thus we conclude that all trajectories but two
approach (2,2) tangent to the straight line with slope
.37. If $c_1 = 0$, we see that there are exactly two
($c_2 > 0$ and $c_2 < 0$) trajectories that lie on the straight
line with slope -1.37.

In similar fashion, we find the linearized solutions near
(1,0) and (0,5/3) to be, respectively,
$$\begin{pmatrix} u \\ v \end{pmatrix} = c_1 \begin{pmatrix} 1 \\ 0 \end{pmatrix} e^{-t} + c_2 \begin{pmatrix} 1 \\ 15/2 \end{pmatrix} e^{11t/4} \quad \text{and}$$
$$\begin{pmatrix} u \\ v \end{pmatrix} = c_1 \begin{pmatrix} 0 \\ 1 \end{pmatrix} e^{-5t/2} + c_2 \begin{pmatrix} 1 \\ 5/52 \end{pmatrix} e^{11t/6},$$

which, along with the
above analysis, yields
the sketch shown:

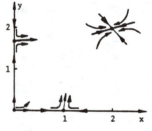

6e. From the above sketch, it appears that $(x,y) \to (2,2)$ as
$t \to \infty$ as long as (x,y) starts in the first quadrant.
To ascertain this, we need to prove that x and y cannot
become unbounded as $t \to \infty$. From the given system, we
can observe that, since $x > 0$ and $y > 0$, that dx/dt and
dy/dt have the same sign as the quantities $1 - x + y/2$
and $5/2 - 3y/2 + x/4$ respectively. If we set these
quantities equal to zero we get the straight lines
$y = 2x - 2$ and $y = x/6 + 5/3$, which divide the first
quadrant into the four
sectors shown. The signs
of x' and y' are indicated,
from which it can be
concluded that x and y
must remain bounded [and
in fact approach (2,2)] as
$t \to \infty$. The discussion
leading up to Fig.9.4.8
is also useful here.

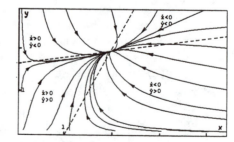

8a. Setting the right sides of the equations equal to zero
     gives the critical points $(0,0)$, $(0, \varepsilon_2/\sigma_2)$, $(\varepsilon_1/\sigma_1, 0)$,
     and possibly
     $([\varepsilon_1\sigma_2 - \varepsilon_2\alpha_1]/[\sigma_1\sigma_2 - \alpha_1\alpha_2], [\varepsilon_2\sigma_1 - \varepsilon_1\alpha_2]/[\sigma_1\sigma_2 - \alpha_1\alpha_2])$.
     (The last point can be obtained from Eq.(37) also). The
     conditions $\varepsilon_2/\alpha_2 > \varepsilon_1/\sigma_1$ and $\varepsilon_2/\sigma_2 > \varepsilon_1/\alpha_1$ imply that
     $\varepsilon_2\alpha_1 - \varepsilon_1\alpha_2 > 0$ and $\varepsilon_1\sigma_2 - \varepsilon_2\alpha_1 < 0$. Thus either the x
     coordinate or the y coordinate of the last critical point
     is negative so a mixed state is not possible. The
     linearized system for $(0,0)$ is $x' = \varepsilon_1 x$ and $y' = \varepsilon_2 y$ and
     thus $(0,0)$ is an unstable equilibrium point. Similarly,
     it can be shown [by linearizing the given system or by
     using Eq.(36)] that $(0, \varepsilon_2/\sigma_2)$ is an asymptotically
     stable critical point and that $(\varepsilon_1\sigma_1, 0)$ is an unstable
     critical point. Thus the fish represented by (redear)
     survive.

8b. The conditions $\varepsilon_1/\sigma_1 > \varepsilon_2/\alpha_2$ and $\varepsilon_1/\alpha_1 > \varepsilon_2/\sigma_2$ imply that
     $\varepsilon_2\sigma_1 - \varepsilon_1\alpha_2 < 0$ and $\varepsilon_1\sigma_2 - \varepsilon_2\alpha_1 > 0$ so again one of the
     coordinates of the fourth point in 8a. is negative and
     hence a mixed state is not possible. An analysis similar
     to that in part(a) shows that $(0,0)$ and $(0,\varepsilon_2/\sigma_2)$ are
     unstable while $(\varepsilon_1/\sigma_1,0)$ is stable. Hence the bluegill
     (represented by x) survive in this case.

9b. If B is reduced, it is clear from the answer to part(a)
     that X is reduced and Y is increased. To determine
     whether the bluegill will die out, we give an intuitive
     argument which can be confirmed by doing the analysis.
     Note that $B/\gamma_1 = \varepsilon_1/\alpha_1 > \varepsilon_2/\sigma_2 = R$ and
     $R/\gamma_2 = \varepsilon_2/\alpha_2 > \varepsilon_1/\sigma_1 = B$ so that the graph of the lines
     $1 - x/B - \gamma_1 y/B = 0$ and $1 - y/R - \gamma_2 x/R = 0$ must appear
     as indicated in the figure,
     where critical points
     are inidcated by heavy
     dots. As B is decreased,
     X decreases, Y increases
     (as indicated above) and
     the point of intersection
     moves closer to $(0,R)$. If
     $B/\gamma_1 < R$ coexistence is not
     possible, and the only
     critical points are $(0,0),(0,R)$ and $(B,0)$.
     It can be shown that $(0,0)$ and $(B,0)$ are unstable and
     $(0,R)$ is asymptotically stable. Hence we conlcude, when
     coexistence is no longer possible, that $x \to 0$ and $y \to R$
     and thus the bluegill population will die out.

3a. We have x = 0 or (1 - .5x - .5y) = 0 and y = 0 or
(-.25 + .5x) = 0 and thus we have three critical points:
(0,0), (2,0) and (1/2,3/2).

3b. For (0,0) the linear system is dx/dt = x and

dy/dt = -.25y and hence $A = \begin{pmatrix} 1 & 0 \\ 0 & -1/4 \end{pmatrix}$ which has

eigenvalues $r_1 = 1$ and $r_2 = -1/4$ and corresponding

eigenvectors $\begin{pmatrix} 1 \\ 0 \end{pmatrix}$ and $\begin{pmatrix} 0 \\ 1 \end{pmatrix}$. Thus (0,0) is an unstable

saddle point.
For (2,0), we let x = 2 + u and y = v in the given

equations and obtain $\dfrac{du}{dt} = -(u+v) - \dfrac{1}{2}u(u+v)$ and

$\dfrac{dv}{dt} = \dfrac{3}{4}v + \dfrac{1}{2}uv$. The linear portion of this has matrix

$A = \begin{pmatrix} -1 & -1 \\ 0 & 3/4 \end{pmatrix}$, which has the eigenvalues $r_1 = -1$,

$r_2 = 3/4$ and corresponding eigenvectors $\begin{pmatrix} 1 \\ 0 \end{pmatrix}$ and $\begin{pmatrix} -4 \\ 7 \end{pmatrix}$.

Thus (2,0) is also an unstable saddle point.

For $\left(\dfrac{1}{2}, \dfrac{3}{2} \right)$ we let x = 1/2 + u and y = 3/2 + v in the

given equations, which yields $\dfrac{du}{dt} = -\dfrac{1}{4}u - \dfrac{1}{4}v, \dfrac{dv}{dt} = \dfrac{3}{4}u$

as the linear portion. Thus $A = \begin{pmatrix} -\dfrac{1}{4} & -\dfrac{1}{4} \\ \dfrac{3}{4} & 0 \end{pmatrix}$, which has

eigenvalues $r_{1,2} = (-1 \pm \sqrt{11}\,i)/8$. Thus $\left(\dfrac{1}{2}, \dfrac{3}{2} \right)$ is an

asymptotically stable spiral point since the eigenvalues
are complex with negative real part. Using
$r_1 = (-1 + \sqrt{11}\,i)/8$ we find that one eigenvector is

$$\begin{pmatrix} -2 \\ 1 + \sqrt{11}\,i \end{pmatrix}$$ and by Section 7.6 the second eigenvector is $$\begin{pmatrix} -2 \\ 1 - \sqrt{11}\,i \end{pmatrix}.$$

3d.

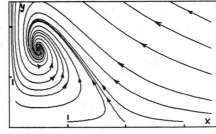

3e. For (x,y) above the line $x + y = 2$ we see that $x' < 0$ and thus x must remain bounded. For (x,y) to the right of $x = 1/2$, $y' > 0$ so it appears that y could grow large asymptotic to $x = $ constant. However, this implies a contradiction ($x = $ constant implies $x' = 0$, but as y gets larger, x' gets increasingly negative) and hence we conclude y must remain bounded and hence $(x,y) \rightarrow (1/2, 3/2)$ as $t \rightarrow \infty$, again assuming they start in the first quadrant.

7a. The amplitude ratio is $(cK/\gamma)/(\sqrt{ac}\,K/\alpha) = \alpha\sqrt{c}/\gamma\sqrt{a}$.

7b. In this case $\alpha = .5$, $a = 1$, $\gamma = .25$ and $c = .75$, so the ratio is $.5\sqrt{.75}/.25\sqrt{1} = 2\sqrt{.75} = \sqrt{3}$.

7c. A rough measurement of the amplitudes is 2.5 and 1.4 and thus the ratio is approximately 1.8. In this case the linear approximation is a good predictor.

10. The presence of a trapping company actually would require a modification of the equations, either by altering the coefficients or by including nonhomogeneous terms on the right sides of the D.E. The effects of indiscreminate trapping could decrease the populations of both rabbits and fox significantly or decrease the fox population which could possibly lead to a large increase in the rabbit population. Over the long run it makes sense for a trapping company to operate in such a way that a consistent supply of pelts is available and to distrub the predator-prey system as little as possible. Thus, the company should trap fox only when their population is increasing, trap rabbits only when their population is increasing, trap rabbits and fox only during the time

when both their populations are increasing, and trap
neither during the time both their populations are
decreasing. In this way the trapping company can have a
moderating effect on the population fluctuations, keeping
the trajectory close to the center.

12. The critical points of the system are the solutions of
the algebraic equations $H(a - \sigma H - \alpha P) = 0$, and
$P(-c + \gamma H) = 0$. the critical points are $H = 0$, $P = 0$;
$H = a/\sigma$, $P = 0$; and $H = c/\gamma$, $P = a/\alpha - c\sigma/\alpha\gamma = \sigma A/\alpha$
where $A = a/\sigma - c/\gamma > 0$.
To study the critical point $(0,0)$ we discard the
nonlinear terms in the system of D.E. to obtain the
corresponding linear system $dH/dt = aH$, $dP/dt = -cP$. The
characteristic equation is $r^2 - (a+c)r - ac = 0$ so
$r_1 = a$, $r_2 = -c$. Thus the critical point $(0,0)$ is an
unstable saddle point.
To study the critical point $(a/\sigma, 0)$ we let $H = (a/\sigma) + u$,
$P = 0 + v$ and substitute in the D.E. to obtain the almost
linear system $du/dt = -au - (a\alpha/\sigma)v - \sigma u^2 - \alpha uv$,
$dv/dt = \gamma Av + \gamma uv$. The corresponding linear system is
$du/dt = -au - (a\alpha/\sigma)v$, $dv/dt = \gamma Av$. The characteristic
equation is $r^2 + (a - \gamma A)r - a\gamma A = 0$ so $r_1 = -a$, $r_2 = \gamma A$.
Thus the critical point $(a/\sigma, 0)$ is an unstable saddle
point.
To study the critical point $(c/\gamma, \sigma A/\alpha)$ we let
$H = (c/\gamma) + u$, $P = (\sigma A/\alpha) + v$ and substitute in the D.E.
to obtain the almost linear system

$$\frac{du}{dt} = -(c\sigma/\gamma)u - (ac/\gamma)v - \sigma u^2 - \alpha uv$$

$$\frac{dv}{dt} = (\sigma A\gamma/\alpha)u + \gamma uv$$

The corresponding linear system is
$du/dt = -(c\sigma/\gamma)u - (\alpha c/\gamma)v$, $dv/dt = (\sigma A\gamma/\alpha)u$. The
characteristic equation is $r^2 + (c\sigma/\gamma)r + c\sigma A = 0$, so
$r_1, r_2 = [-(c\sigma/\gamma) \pm \sqrt{(c\sigma/\gamma)^2 - 4c\sigma A}]/2$. Thus, depending
on the sign of the discriminant we have that $(c/\gamma, \sigma A/\alpha)$
is either an asymptotically stable spiral point or an
asymptotically stable improper node. Thus for nonzero
initial data $(H,P) \to (c/\gamma, \sigma A/\alpha)$ as $t \to \infty$.

Section 9.6, Page 489

1. Assuming that $V(x,y) = ax^2 + cy^2$ we find $V_x(x,y) = 2ax$,

$V_y = 2cy$ and thus Eq.(7) yields $\dot{V}(x,y) = 2ax(-x^3 + xy^2) +$

$2cy(-2x^2y - y^3) = -[2ax^4 + 2(2c-a)x^2y^2 + 2cy^4]$. If we choose a and c to be any positive real numbers with

$2c > a$, then $\dot{V}$ is a negative definite. Also, V is positive definite by Theorem 9.6.4. Thus by Theorem 9.6.1 the origin is an asymptotically stable critical point.

3. Assuming the same form for V(x,y) as in Problem 1, we have

$$\dot{V}(x,y) = 2ax(-x^3 + 2y^3) + 2cy(-2xy^2) = -2ax^4 + 4(a-c)xy^3.$$

If we choose a = c > 0, then $\dot{V}(x,y) = -2ax^4 \leq 0$ in any

neighborhood containing the origin and thus $\dot{V}$ is negative semidefinite and V is positive definite. Theorem 9.6.1 then concludes that the origin is a stable critical point. Note that the origin may still be asymptotically stable, however, the V(x,y) used here is not sufficient to prove that.

6a. The correct system is dx/dt = y and dy/dt = -g(x). Since g(0) = 0, we conclude that (0,0) is a critical point.

6b. From the given conditions, the graph of g must be positive for 0 < x < k and negative for -k < x < 0. Thus if   0 < x < k   then $\int_0^x g(s)\,ds > 0$,

if   -k < x < 0   then $\int_0^x g(s)\,ds = -\int_x^0 g(s)\,dx > 0$.

Since V(0,0) = 0 it follows that $V(x,y) = y^2/2 + \int_0^x g(s)\,ds$

is positive definite for -k < x < k, $-\infty < y < \infty$. Next,

we consider $\dot{V}(x,y) = y[-g(x)] + g(x)y = 0$. Since $\dot{V}(x,y)$ is never positive, we may conclude that it is negative semidefinite and hence by Theorem 9.6.1 (0,0) is at least a stable critical point.

7b. V is positive definite by Theorem 9.6.4. Since $V_x(x,y) = 2x$, $V_y(x,y) = 2y$, we obtain

$\dot{V}(x,y) = 2xy - 2y^2 - 2y\sin x = 2y[-y + (x - \sin x)]$. If

x < 0, then $\dot{V}(x,y) < 0$ for all y > 0. If x > 0, choose y

so that 0 < y < x - sin x. Then $\dot{V}(x,y) > 0$. Hence V is not a Liapunov function. Since V(0,0) = 0, 1 - cos x > 0 for 0 < |x| < 2π and $y^2 > 0$ for y ≠ 0, it follows that V(x,y) is positive definite in a neighborhood of the origin. Next $V_x(x,y) = \sin x$, $V_y(x,y) = y$, so

$V(x,y) = (\sin x)(y) + y(-y - \sin x) = -y^2$. Hence $\dot{V}$ is
negative semidefinite and $(0,0)$ is a stable critical
point by Theorem 9.6.1.

7d. $V(x,y) = (x+y)^2/2 + x^2 + y^2/2 = 3x^2/2 + xy + y^2$ is
positive definite by Theorem 9.6.4. Next
$V_x(x,y) = 3x + y$, $V_y(x,y) = x + 2y$ so

$$
\begin{aligned}
\dot{V}(x,y) &= (3x+y)y - (x+2y)(y+\sin x) \\
&= 2xy - y^2 - (x+2y)\sin x \\
&= 2xy - y^2 - (x+2y)(x - \alpha x^3/6) \\
&= -x^2 - y^2 + \alpha(x+2y)x^3/6 \\
&= -r^2 + \alpha r^4 (\cos\theta + 2\sin\theta)(\cos^3\theta)/6 < -r^2 + r^4/2 \\
&= -r^2(1-r^2/2).
\end{aligned}
$$

Thus $\dot{V}$ is negative definite for
$r < \sqrt{2}$. From Theorem 9.6.1 it follows that the origin
is an asymptotically stable critical point.

8. Let $x = u$ and $y = du/dt$ to obtain the system $dx/dt = y$
and $dy/dt = -c(x)y - g(x)$. Now consider
$V(x,y) = y^2/2 + \int_0^x g(s)\,ds$.

10b. Since $V_x(x,y) = 2Ax + By$, $V_y(x,y) = Bx + 2Cy$, we have

$\dot{V}(x,y) = (2Ax + By)(ax + by) + (Bx + 2Cy)(cx + dy) =$
$(2Aa + Bc)x^2 + [2(Ab + Cc) + B(a+d)]xy + (2Cd + Bb)y^2$.
We choose A, B, and C so that $2Aa + Bc = -1$,
$2(Ab + Cc) + B(a+d) = 0$, $2Cd + Bb = -1$. The first and
third equations give us A and C in terms of B,
respectively. We substitute in the second equation to
find B and then calculate A and C. The result is given in
the text.

10c. Since $ad - bc > 0$ and $a + d < 0$, we see that $\Delta < 0$ and
so $A > 0$. Using the expressions for A, B, and C found in
part (b) we obtain
$$
\begin{aligned}
(4AC-B^2)\Delta^2 &= [c^2+d^2 + (ad-bc)][a^2+b^2 + (ad-bc)] - (bd+ac)^2 \\
&= (a^2+b^2+c^2+d^2)(ad-bc) + (a^2+b^2)(c^2+d^2) + (ad-bc)^2 \\
&\qquad\qquad\qquad\qquad\qquad\qquad\qquad\qquad - (bd+ac)^2 \\
&= (a^2+b^2+c^2+d^2)(ad-bc) + 2(ad-bc)^2.
\end{aligned}
$$
Since $ad - bc > 0$ it follows that $4AC - B^2 > 0$.

11b. Substituting $x = r\cos\theta$, $y = r\sin\theta$ we find that

$\dot{V}[x(r,\theta), y(r,\theta)] = -r^2 + r(2A\cos\theta + B\sin\theta)F_1[x(r,\theta), y(r,\theta)]$
$+ r(B\cos\theta + 2C\sin\theta)G_1[x(r,\theta), y(r,\theta)]$. Now we make use
of the facts that (1) there exists an M such that

$|2A| \le M$, $|B| \le M$, and $|2C| \le M$; and (2) given any
$\varepsilon > 0$ there exists a circle $r = R$ such that
$|F_1(x,y)| < \varepsilon r$ and $|G_1(x,y)| < \varepsilon r$ for $0 \le r < R$.   We have
$|2A\cos\theta + B\sin\theta| \le 2M$ and $|B\cos\theta + 2C\sin\theta| \le 2M$.   Hence

$$\dot{V}[x(r,\theta), \ y(r,\theta)] \le -r^2 + 2Mr(\varepsilon r) + 2Mr(\varepsilon r) = -r^2(1 - 4M\varepsilon).$$

If we choose $\varepsilon = M/8$ we obtain $\dot{V}[x(r,\theta), \ y(r,\theta)] \le -r^2/2$

for $0 \le r < R$.   Hence $\dot{V}$ is negative definite in $0 \le r < R$
and from Problem 10c V is positive definite and thus V is
a Liapunov function for the almost linear system.

Section 9.7, Page 500

1. $r = 1$, $\theta = t + t_0$ is a periodic solution.   If $r < 1$, then
    $dr/dt > 0$, and the direction of motion on a trajectory is
    outward.   If $r > 1$, then the direction of motion is
    inward.   It follows that the periodic solution $r = 1$,
    $\theta = t + t_0$ is a stable limit cycle.

2. $r = 1$, $\theta = -t + t_0$ is a periodic solution.   If $r < 1$,
    then $dr/dt > 0$, and the direction of motion on a
    trajectory is outward.   If $r > 1$, the $dr/dt > 0$, and the
    direction of motion is still outward.   It follows that
    the solution $r = 1$, $\theta = -t + t_0$ is a semistable limit
    cycle.

4. $r = 1$, $\theta = -t + t_0$ and $r = 2$, $\theta = -t + t_0$ are periodic
    solutions.   If $r < 1$, then $dr/dt < 0$, and the direction
    of motion on a trajectory is inward.   If $1 < r < 2$, then
    $dr/dt > 0$, and the direction of motion is outward.
    Similarly, if $r > 2$, the direction of motion is inward.
    It follows that the periodic solution $r = 1$,
    $\theta = -t + t_0$ is unstable and the periodic solution $r = 2$,
    $\theta = -t + t_0$ is a stable limit cycle.

7. Differentiating x and y with respect to t we find that
    $dx/dt = (dr/dt)\cos\theta - (r\sin\theta)d\theta/dt$ and
    $dy/dt = (dr/dt)\sin\theta + (r\cos\theta)d\theta/dt$.   Hence
    $ydx/dt - xdy/dt = (r\sin\theta\cos\theta)dr/dt - (r^2\sin^2\theta)d\theta/dt -$
    $\qquad\qquad\qquad\qquad (r\cos\theta\sin\theta)dr/dt - (r^2\cos\theta)d\theta/dt$
    $\qquad\qquad\qquad = - r^2 d\theta/dt.$

8a. Multiplying the first equation by x and the second by y
    and adding yields $xdx/dt + ydy/dt = (x^2+y^2)f(r)/r$, or
    $rdr/dt = rf(r)$ and thus $dr/dt = rf(r)$.   To obtain an

equation for θ multiply the first equation by y, the
second by x and substract to obtain
$ydx/dt - xdy/dt = -x^2-y^2$, or $-r^2d\theta/dt = -r^2$, using the
results of Problem 7. Thus $d\theta/dt = 1$. It follows that
periodic solutions are given by $r = c$, $\theta = t + t_0$ where
$f(c) = 0$. Since $\theta = t + t_0$ the motion is
counterclockwise.

8b. First note that $f(r) = r(r-2)^2(r-3)(r-1)$. Thus $r = 1$,
$\theta = t + t_0$; $r = 2$, $\theta = t + t_0$; and $r = 3$, $\theta = t + t_0$ are
periodic solutions. If $r < 1$, then $dr/dt > 0$, and the
direction of motion on a trajectory is outward. If
$1 < r < 2$, then $dr/dt < 0$ and the direction of motion is
inward. Thus the periodic solution $r = 1$, $\theta = t + t_0$ is
a stable limit cycle. If $2 < r < 3$, then $dr/dt < 0$, and
the direction of motion is inward. Thus the periodic
solution $r = 2$, $\theta = t + t_0$ is a semistable limit cycle.
If $r > 3$, then $dr/dt > 0$, and the direction of motion is
outward. Thus the periodic solution $r = 3$, $\theta = t + t_0$ is
unstable.

9.   Setting $x = r\cos\theta$, $y = r\sin\theta$ and using the techniques of
Problem 8 the equations transform to $dr/dt = r^2 - 2$,
$d\theta/dt = -1$. This system has a periodic solution $r = \sqrt{2}$,
$\theta = -t + t_0$. If $r < \sqrt{2}$, then $dr/dt < 0$, and the
direction of motion along a trajectory is inward. If
$r > \sqrt{2}$, then $dr/dt > 0$, and the direction of motion is
outward. Thus the periodic solution $r = \sqrt{2}$, $\theta = -t + t_0$
is unstable.

11. If $F(x,y) = x+y+x^3-y^3$, $G(x,y) = -x+2y+x^2y+y^3/3$, then
$F_x(x,y) + G_y(x,y) = 1+3x^2+2+x^2+y^2 = 3+4x^2+y^2$. Since the
conditions of Theorem 9.7.2 are satisfied for all x and
y, it follows that the system has no periodic nonconstant
solution.

13. Since $x = \phi(t)$, $y = \psi(t)$ is a solution of Eqs.(15), we
have $d\phi/dt = F[\phi(t),\psi(t)]$, $d\psi/dt = G[\phi(t),\psi(t)]$. Hence
on the curve C,
$F(x,y)dy - G(x,y)dx = \phi'(t)\psi'(t)dt -\psi'(t)\phi'(t)dt = 0$. It
follows that the line integral around C is zero.
However, if $F_x + G_y$ has the same sign throughout D, then
the double integral cannot be zero. This gives a
contradiction. Thus either the solution of Eqs.(15) is
not periodic or if it is, it cannot lie entirely in D.

Section 9.8, Page 508

1b. For $\lambda = \lambda_3 = (-11 + \alpha)/2$, where $\alpha = \sqrt{81+40r}$, we have

$$\begin{pmatrix} -10+(11-\alpha)/2 & 10 & 0 \\ r & -1+(11-\alpha)/2 & 0 \\ 0 & 0 & -8/3+(11-\alpha)/2 \end{pmatrix}\begin{pmatrix} \xi_1 \\ \xi_2 \\ \xi_3 \end{pmatrix} = \begin{pmatrix} 0 \\ 0 \\ 0 \end{pmatrix}.$$

The last line implies $\xi_3 = 0$ and multiplying the first line by

$(-9+\alpha)/2$ we obtain $\begin{pmatrix} (81-\alpha^2)/4 & 10(-9+\alpha)/2 \\ r & (9-\alpha)/2 \end{pmatrix}\begin{pmatrix} \xi_1 \\ \xi_2 \end{pmatrix} = \begin{pmatrix} 0 \\ 0 \end{pmatrix}.$

Substituting $\alpha^2 = 81+40r$ we have

$$\begin{pmatrix} -10r & -10(9-\alpha)/2 \\ r & (9-\alpha)/2 \end{pmatrix}\begin{pmatrix} \xi_1 \\ \xi_2 \end{pmatrix} = \begin{pmatrix} 0 \\ 0 \end{pmatrix}.$$

Thus $\xi^{(3)} = \begin{pmatrix} 9 - \sqrt{81+40r} \\ -2r \\ 0 \end{pmatrix}$, which is proportional to the

answer given in the text.

2a. The calculations are somewhat simplified if you let
$x = \beta + u$, $y = \beta + v$, and $z = (r-1)+w$, where
$\beta = \sqrt{8(r-1)/3}$.

2b. Eq.(8) is found by evaluating $\begin{vmatrix} -10-\lambda & 10 & 0 \\ 1 & -1-\lambda & -\beta \\ \beta & \beta & -8/3-\lambda \end{vmatrix} = 0.$

3b. If r_1, r_2, r_3 are the three roots of a cubic polynomial,
then the polynomial can be factored as
$(x-r_1)(x-r_2)(x-r_3)$. Expanding this and equating to the
given polynomial we have $A = -(r_1+r_2+r_3)$,
$B = r_1r_2 + r_1r_3 + r_2r_3$ and $C = -r_1r_2r_3$. We are interested
in the case when the real part of the complex conjugate
roots changes sign. Thus let $r_2 = \alpha+i\beta$ and $r_3 = \alpha-i\beta$,
which yields
$A = -(r_1+2\alpha)$, $B = 2\alpha r_1 + \alpha^2 + \beta^2$ and $C = r_1(\alpha^2+\beta^2)$.
Hence, if $AB = C$, we have
$-(r_1+2\alpha)(2\alpha r_1+\alpha^2+\beta^2) = -r_1(\alpha^2+\beta^2)$ or
$-2\alpha[r_1^2 + 2\alpha r_1 + (\alpha^2+\beta^2)] = 0$ or
$-2\alpha[(r_1+\alpha)^2+\beta^2] = 0$. Since the square bracket
term is positive, we conclude that if $AB = C$, then
$\alpha = 0$. That is, the conjugate complex roots are pure

imaginary. Note that the converse is also true. That
is, if the conjugate complex roots are pure imaginary
then AB = C.

4. We have

$$\dot{V} = 2x[\sigma(-x+y)] + 2\sigma y[rx-y-xz] + 2\sigma z[-bz+xy]$$
$$= -2\sigma x^2 + 2\sigma xy + 2\sigma rxy - 2\sigma y^2 - 2\sigma bz^2$$
$$= 2\sigma\{-[x^2-(r+1)xy+y^2]-z^2\}.$$ For r < 1, the term in the

square brackets remains positive for all values of x and

y, and thus $\dot{V}$ is negative definite. Thus, by the
extension of Theorem 9.6.1 to three equations, we
conclude that the origin is an asymptotically stable
critical point.

5a. $V = rx^2 + \sigma y^2 + \sigma(z-2r)^2 = c > 0$ yields

$$\frac{dv}{dt} = 2rx[\sigma(-x+y)] + 2\sigma y(rx-y-xz) + 2\sigma(z-2r)(-bz+xy)$$ when

Eqs.(1) are substituted in. Thus

$$\dot{V} = -2\sigma[rx^2 + y^2 + b(z^2 - 2rz)]$$
$$= -2\sigma[rx^2 + y^2 + b(z-r)^2 - br^2].$$

5b. From the proof of Theorem 9.6.1, we find that we need to

show that $\dot{V}$, as found in part a, is always negative as it
crosses V(x,y,z) = c. (Actually, we need to use the
extension of Theorem 9.6.1 to three equations, but the
proof is very similar using the vector calculus
approach.) From part a we see that

$\dot{V} < 0$ if $rx^2 + y^2 + b(z-r)^2 > br^2$, which holds if (x,y,z)

lies outside the ellipsoid $\dfrac{x^2}{br} + \dfrac{y^2}{br^2} + \dfrac{(z-r)^2}{r^2} = 1.$ (i)

Thus we need to choose c such that V = c lies outside
Eq.(i). Writing V = c in the form of Eq.(i) we obtain

the ellipsoid $\dfrac{x^2}{c/r} + \dfrac{y^2}{c/\sigma} + \dfrac{(z-2r)^2}{c/\sigma} = 1.$ (ii) Now let

$M = \max(\sqrt{br}, r\sqrt{br}, r)$, then the ellipsoid (i) is

contained inside the sphere S1: $\dfrac{x^2}{M^2} + \dfrac{y^2}{M^2} + \dfrac{(z-r)^2}{M^2} = 1.$

Let S2 be a sphere centered at (0,0,2r) with radius

M+r: $\dfrac{x^2}{(M+r)^2} + \dfrac{y^2}{(M+r)^2} + \dfrac{(z-2r)}{(M+r^2)} = 1,$ then S1 is contained

in S2. Thus, if we choose c, in Eq.(ii), such that

$\dfrac{c}{r} > (M+r)^2$ and $\dfrac{c}{\sigma} > (M+r)^2$, then $\dot{V} < 0$ as the trajectory crosses $V(x,y,z) = c$. Note that this is a sufficient condition and there may be many other "better" choices using different techniques.

8b. Several cases are shown. Results may vary, particularly for $r = 24$, due to the closeness of r to $r_3 \cong 24.06$.

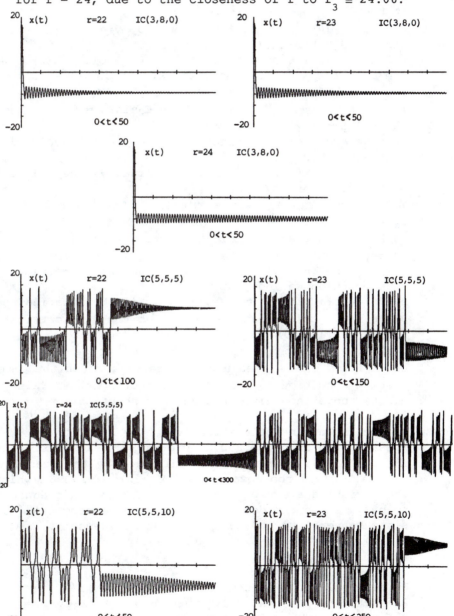

CHAPTER 10

5. Following the procedures of Eqs.(5) through (8), we set
    $u(x,y) = X(x)T(t)$ in the P.D.E. to obtain $X''T = 4XT'$, or
    $X''/X = 4T'/T$, which must be a constant. As shown in the
    text this separation constant must be $-\lambda^2$ and thus
    $X'' + \lambda^2 X = 0$ and $T' + (\lambda^2/4)T = 0$. Now
    $u(0,t) = X(0)T(t) = 0$, for all $t > 0$, yields $X(0) = 0$, as
    discussed after Eq.(11) and similarly
    $u(2,t) = X(2)T(t) = 0$, for all $t > 0$, implies $X(2) = 0$.
    The D.E. for X has the solution $X(x) = C_1\cos\lambda x + C_2\sin\lambda x$
    and $X(0) = 0$ yields $C_1 = 0$. Setting $x = 2$ in the
    remaining form of X yields $X(2) = C_2\sin2\lambda = 0$, which has
    the solutions $2\lambda = n\pi$ or $\lambda = n\pi/2$, $n = 1,2,\ldots$. Note
    that we exclude $n = 0$ since then $\lambda = 0$ which would yield
    $X(x) = 0$, which is unacceptable. Hence
    $X(x) = \sin(n\pi x/2)$, $n = 1,2,\ldots$. Finally, the solution
    of the D.E. for T yields
    $T(t) = \exp(-\lambda^2 t/4) = \exp(-n^2\pi^2 t/16)$. Thus we have found
    $u_n(x,t) = \exp(-n^2\pi^2 t/16)\sin(n\pi x/2)$. Setting $t = 0$ in
    this last expression indicates that $u_n(x,0)$ has, for the
    correct choices of n, the same form as the terms in
    $u(x,0)$, the initial condition. Using the principle of
    superposition we know that
    $u(x,t) = c_1 u_1(x,t) + c_2 u_2(x,t) + c_4 u_4(x,t)$ satisfies the
    P.D.E. and the B.C. and hence we let $t = 0$ to obtain
    $u(x,0) = c_1 u_1(x,0) + c_2 u_2(x,0) + c_4 u_4(x,0) =$
    $c_1\sin n\pi x/2 + c_2\sin n\pi x + c_4\sin 2\pi x$. If we choose $c_1 = 2$,
    $c_2 = -1$ and $c_4 = 4$ then $u(x,0)$ here will match the given
    initial condition, and hence substituting these values in
    $u(x,t)$ above then gives the desired solution.

9. We seek solutions of the form $u(x,t) = X(x)T(t)$.
    Substituting into the P.D.E. yields $X''T + X'T' + XT' =$
    $X''T + (X' + X)T' = 0$. Formally dividing by the quantity
    $(X' + X)T$ gives the equation $X''/(X' + X) = -T'/T$ in which
    the variables are separated. In order for this equation
    to be valid on the domain of u it is necessary that both
    sides be equal to the same constant λ. Hence
    $X''/(X' + X) = -T'/T = \lambda$ or equivalently, $X'' - \lambda(X' + X) = 0$
    and $T' + \lambda T = 0$.

11. We seek solutions of the form $u(x,y) = X(x)Y(y)$.
    Substituting into the P.D.E. yields $X''Y + (x+y)XY'' =$
    $X''Y + xXY'' + yXY'' = 0$. Formally dividing by XY yields

$X''/X + xY''/Y + yY''/Y = 0$. From this equation we see that the presence of the independent variable x multiplying the term u_{yy} in the original equation leads to the term xY''/Y when we attempt to separate the variables. It follows that the argument for a separation constant does not carry through and we cannot replace the P.D.E. by two O.D.E.

14. Applying the chain rule to partial differentiation of u with respect to x we see that $u_x = u_\xi \xi_x = u_\xi (1/\ell)$ and $u_{xx} = u_{\xi\xi}(1/\ell)^2$. Substituting $u_{\xi\xi}/\ell^2$ for u_{xx} in the heat equation gives $\alpha^2 u_{\xi\xi}/\ell^2 = u_t$ or $u_{\xi\xi} = (\ell^2/\alpha^2)u_t$. A similar argument for the t substitution then yields the desired P.D.E.

16. Substituting $u(x,y,t) = X(x)Y(y)T(t)$ in the P.D.E. yields $\alpha^2(X''YT + XY''T) = XYT'$, which is equivalent to $X''/X + Y''/Y = T'/\alpha^2 T$. By keeping the independent variables x and y fixed and varying t we see that $T'/\alpha^2 T$ must equal some constant σ_1 since the left side of the equation is fixed. Hence, $X''/X + Y''/Y = T'/\alpha^2 T = \sigma$, or $X''/X = \sigma_1 - Y''/Y$ and $T' - \sigma_1\alpha^2 T = 0$. By keeping x fixed and varying y in the equation involving X and Y we see that $\sigma_1 - Y''/Y$ must equal some constant σ_2 since the left side of the equation is fixed. Hence, $X''/X = \sigma_1 - Y''/Y = \sigma_2$ so $X'' - \sigma_2 X = 0$ and $Y'' - (\sigma_1 - \sigma_2)Y = 0$. If we assume homogeneous B.C. and require that solutions be bounded as $t \to \infty$, then it can be shown by an argument similar to the one in the text that σ_1 and σ_2 must be negative constants, $\sigma_1 = -\lambda^2$, $\sigma_2 = -\mu^2$. Then the O.D.E. become $X'' + \mu^2 X = 0$, $Y'' + (\lambda^2 - \mu^2)Y = 0$, and $T' + \alpha^2\lambda^2 T = 0$.

Section 10.2, Page 526

3. We look for values of T for which $\sinh 2(x+T) = \sinh 2x$ for all x. Expanding the left side of this equation gives $\sinh 2x \cosh 2T + \cosh 2x \sinh 2T = \sinh 2x$, which will be satisfied for all x if we can choose T so that $\cosh 2T = 1$ and $\sinh 2T = 0$. The only value of T satisfying these constraints is $T = 0$. Since T is not positive we conclude that the function $\sinh 2x$ is not periodic.

4. We look for values of T for which $\tan \pi(x+T) = \tan \pi x$. Expanding the left side gives

$\tan\pi(x+T) = (\tan\pi x + \tan\pi T)/(1-\tan\pi x\tan\pi T)$ which is equal to $\tan\pi x$ only for $\tan\pi T = 0$. The only positive solutions of this last equation are $T = 1,2,3...$ and hence $\tan\pi x$ is periodic with fundamental period $T = 1$.

11a. First we have $\int_T^{a+T} g(x)\,dx = \int_0^a g(s)\,ds$ by letting $x = s + T$ in the left integral. Now, if $0 \le a \le T$, then from elementary calculus we know that $\int_a^{a+T} g(x)\,dx =$

$\int_a^T g(x)\,dx + \int_T^{a+T} g(x)\,dx = \int_a^T g(x)\,dx + \int_0^a g(x)\,dx$ using the equality derived above. This last sum is $\int_0^T g(x)\,dx$ and thus we have the desired result.

It is recommended that the given function and its periodic extension be graphed for several periods for each of the Problems 14 through 23. In many of these problems it is necessary to use integration by parts to evaluate the coefficients, although all the details will be not shown here.

14. The function represents a sawtooth wave, as indicated in the figure, and its periodic with period 2ℓ. Thus the Fourier series is of the form

$$f(x) = a_0/2 + \sum_{m=1}^{\infty} (a_m\cos m\pi x/\ell + b_m\sin m\pi x/\ell)$$ where the

coefficients are computed from Eqs. (13) and (14).

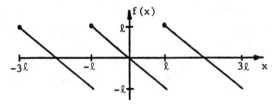

Substituting for $f(x)$ in these equations yields

$a_0 = (1/\ell)\int_{-\ell}^{\ell}(-x)\,dx = 0;\quad a_m = (1/\ell)\int_{-\ell}^{\ell}(-x)\cos(m\pi x/\ell)\,dx = 0,$

$m = 1,2...;$ and

$b_m = (1/\ell)\int_{-\ell}^{\ell}(-x)\sin(m\pi x/\ell)\,dx$

$\qquad = (x/m\pi)\cos(m\pi x/\ell)\Big|_{-\ell}^{\ell} - (1/m\pi)\int_{-\ell}^{\ell}\cos(m\pi x/\ell)\,dx$

$\qquad = (2\ell\cos m\pi)/m\pi = 2\ell(-1)^m/m\pi.$

Substituting these terms in the above Fourier series for $f(x)$ yields the desired answer.

18. The function represents a sawtooth wave, as indicated in
     the figure, and although it is periodic with period 2 we
     see that the fundamental period is actually 1. Thus the

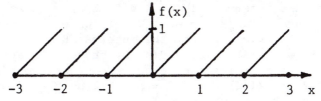

     Fourier series is of the form

     $f(x) = a_0/2 + \sum\limits_{m=1}^{\infty} (a_m\cos2m\pi x + b_m\sin2m\pi x)$ where the

     coefficients are computed from Eqs. (13) and (14) with ℓ
     replaced by 1/2. Using the results of Problem 11c or the
     discussion just prior to Example 1 we may change the
     interval of integration from [-1/2, 1/2] to [0,1].
     Substituting $f(x) = x$ into these equations with the above
     changes yields $a_0 = 2\int_0^1 x\,dx = 1$; $a_m = 2\int_0^1 x\cos2m\pi x\,dx = 0$,

     $m = 1,2,...$; $b_m = 2\int_0^1 x\sin2m\pi x\,dx = -1/m\pi$, $m = 1,2...$.
     Substituting these values in the Fourier series for $f(x)$
     above gives the desired result. Note that exactly the
     same results are obtained if Eqs. (13) and (14) are used
     with $\ell = 1$. In this case
     $b_n = \int_{-1}^0 (x+1)\sin n\pi x\,dx + \int_0^1 x\sin n\pi x\,dx$ which yields
     $b_n = -2/n\pi$ for n even and $b_n = 0$ for n odd.

20. In this case $f(x)$ is periodic of period 2π and thus $\ell = \pi$
     in Eqs. (9), (13,) and (14). The constant a_0 is found to
     be $a_0 = (1/\pi)\int_{-\pi}^0 x\,dx = -\pi/2$ since $f(x)$ is zero on the
     interval $[0,\pi]$. Likewise $a_n = (1/\pi)\int_{-\pi}^0 x\cos nx\,dx =$
     $[1 - (-1)^n]n^2\pi$ using integration by parts. Thus $a_n = 0$
     for n even and $a_n = 2/n^2\pi$ for n odd, which may be written
     as $a_{2n-1} = 2/(2n-1)^2\pi$ since 2n-1 is always an odd number.
     In a similar fashion $b_n = (1/\pi)\int_{-\pi}^0 x\sin nx\,dx = (-1)^{n+1}/n$ and
     thus the desired solution is obtained. Notice that in
     this case both cosine and sine terms appear in the
     Fourier series for the given $f(x)$.

25. The graph of $f(x)$ is shown in Problem 18. We note that
     $f(x)$ is a straight line with a slope of one in any
     interval. Thus $f(x)$ has the form x+b in any interval for

the correct value of b. Since f(x+2) = f(x), we may set
x = -1/2 to obtain f(3/2) = f(-1/2). Noting that 3/2 is
on the interval 1 < x < 2[f(3/2) = 3/2 + b] and that -1/2
is on the interval -1 < x < 0[f(-1/2) = -1/2 + 1], we
conclude that 3/2 + b = -1/2 + 1, or b = -1 for the
interval 1 < x < 2. For the interval 8 < x < 9 we have
f(x+8) = f(x+6) = ... = f(x) by successive applications
of the periodicity condition. Thus for x = 1/2 we have
f(17/2) = f(1/2) or 17/2 + b = 1/2 so b = -8 on the
interval 8 < x < 9.

28. Use the procedure outlined in Problem 25 to find
f(x) = x-2 on the interval 1 < x < 2. Thus, for
instance, $b_n = \int_0^2 f(x)\sin n\pi x\,dx = \int_0^1 x\sin n\pi x\,dx +$
$\int_1^2 (x-2)\sin n\pi x\,dx = 2(-1)^{n+1}/n\pi$. Similar integrations for
a_0 and a_n also yield the same results as in Problem 17.

Section 10.3, Page 533

2. The function to which the series converges is indicated
in the figure and is periodic with period 2π. Note that

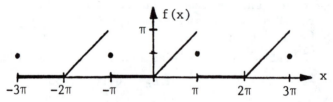

the Fourier series converges to $\pi/2$ at x = $-\pi$, π, etc.,
even though the fuction is defined to be zero there.
This value ($\pi/2$) represents the mean value of the left
and right hand limits at those points. In ($-\pi,0$),
f(x) = 0 and f'(x) = 0 so both f and f' are continuous and
have finite limits as x → $-\pi$ from the right and as
x → 0 from the left. In ($0,\pi$), f(x) = x and f'(x) = 1
and again both f and f' are continuous and have limits as
x → 0 from the right and as x → π from the left. Since
f and f' are piecewise continuous on $[-\pi,\pi)$ the
conditions of the Fourier theorem are satisfied.
Substituting for f(x) in Eqs.(2) and (3) with $\ell = \pi$
yields $a_0 = (1/\pi)\int_0^\pi x\,dx = \pi/2$;
$a_m = (1/\pi)\int_0^\pi x\cos mx\,dx = (\cos m\pi - 1)/\pi m^2 = 0$ for m even and
$= -2/\pi m^2$ for m odd; and

$b_m = (1/\pi) \int_0^\pi x\sin mx \, dx = -(\pi\cos m\pi)/m\pi = (-1)^{m+1}/m,$

$m = 1,2\ldots$.   Substituting these values into Eq.(1) with $\ell = \pi$ yields the desired solution.

3.   The function to which the series converges is shown in the figure and is periodic with a fundamental period of π.   In $(0,\pi)$, $f(x) = \sin^2 x$ and $f'(x) = 2\sin x\cos x = \sin 2x$ so both f and f' are continuous and have finite limits as

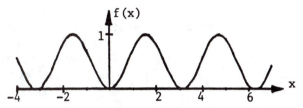

the endpoints of the interval $(0,\pi)$ are approached from within the interval.   Thus f and f' are piecewise continuous (indeed, they are continuous) on the interval $(0,\pi)$ and the conditions of the Fourier theorem are satisfied.   Substituting for $f(x)$ in Eqs.(2) and (3) with $\ell = \pi/2$ and integrating over the interval $0 \le x < \pi$ (or alternatively, $-\pi/2 \le x < \pi/2$) yields

$a_0 = (2/\pi) \int_0^\pi \sin^2 x \, dx = 1;$ $a_m = (2/\pi) \int_0^\pi \sin^2 x\cos 2mx \, dx = -1/2$

if $m = 1$, otherwise $a_m = 0$, $m = 2,3\ldots$;

$b_m = (2/\pi) \int_0^\pi \sin^2 x\sin 2mx \, dx = 0$, $m = 1,2,\ldots$.   Thus the Fourier series for f is $f(x) = 1/2 - (1/2)\cos 2x$. This is not surprising, though, for if we expand $\sin^2 x$ in terms of the trigonometric half angle formula, we also have $\sin^2 x = 1/2 - (1/2)\cos 2x$.

6.   The function to which the series converges is shown in the figure and is periodic of fundamental period 2.   In $[-1,1]$ $f(x)$ and $f'(x) = -2x$ are both continuous and have

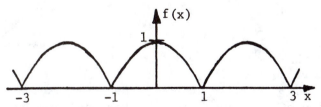

finite limits as the endpoints of the interval are approached from within the interval.   Thus the conditions of the Fourier Theorem are satisfied.   Note that in this case $f(x)$ and its periodic extension are continuous, while $f'(x)$ and its periodic extension are only piecewise

continuous. Substituting for $f(x)$ in Eqs. (2) and (3),
with $\ell = 1$ yields $a_0 = \int_{-1}^{1}(1-x^2)\,dx = 4/3$;

$a_n = \int_{-1}^{1}(1-x^2)\cos n\pi x\,dx = (2/n\pi)\int_{-1}^{1}x\sin n\pi x\,dx =$

$(-2/n^2\pi^2)[x\cos n\pi x\Big|_{-1}^{1} - \int_{-1}^{1}\cos n\pi x\,dx] =$

$4(-1)^{n+1}/n^2\pi^2$; $b_n = \int_{-1}^{1}(1-x^2)\sin n\pi x\,dx = 0$. Substituting
these values into Eq. (1) gives the desired series.

10. The solution to the corresponding homogeneous equation is
found by methods presented in Section 3.4 and is
$y(t) = c_1\cos\omega t + c_2\sin\omega t$. For the nonhomogeneous terms
we use the method of superposition and consider the
sequence of equations $y_n'' + \omega^2 y_n = b_n\sin nt$ for
$n = 1,2,3\ldots$. If $\omega > 0$ is not equal to an integer,
then the particular solution to this last equation has
the form $Y_n = a_n\cos nt + d_n\sin nt$, as previously discussed
in Section 3.6. Substituting this form for Y_n into the
equation and solving, we find $a_n = 0$ and $d_n = b_n/(\omega^2-n^2)$.
Thus the formal general solution of the original
nonhomogeneous D.E. is

$$y(t) = c_1\cos\omega t + c_2\sin\omega t + \sum_{n=1}^{\infty} b_n(\sin nt)/(\omega^2-n^2), \text{ where}$$

we have superimposed all the Y_n terms to obtain the
infinite sum. To evaluete c_1 and c_2 we set $t = 0$ to
obtain $y(0) = c_1 = 0$ and

$$y'(0) = \omega c_2 + \sum_{n=1}^{\infty}nb_n/(\omega^2-n^2) = 0 \text{ where we have formally}$$

differentiated the infinite series term by term and
evaluated it at $t = 0$. (Differentiation of a Fourier
Series has not been justified yet and thus we can only
consider this method a formal solution at this point).
Thus $c_1 = 0$,

$$c_2 = -(1/\omega)\sum_{n=1}^{\infty}nb_n/(\omega^2-n^2), \text{ which when substituted into}$$

the above series yields the desired solution.
If $\omega = m$, a positive integer, then the particular
solution of $y_m'' + m^2 y_m = b_m\sin mt$ has the form
$Y_m = t(a_m\cos mt + d_m\sin mt)$ since $\sin mt$ is a solution of
the related homogeneous D.E. Substituting Y_m into the
D.E. yields $a_m = -b_m/2m$ and $d_m = 0$ and thus the general

solution of the D.E. (when $\omega = m$) is now $y(t) = c_1\cos mt$
$+ c_2\sin mt - b_m t(\cos mt)/2m + \displaystyle\sum_{n=1, n\neq m}^{\infty} b_n(\sin nt)/(\omega^2-n^2)$.   To

evaluate c_1 and c_2 we set $y(0) = c_1 = 0$ and

$y'(0) = c_2 m - b_m/2m + \displaystyle\sum_{n=1, n\neq m}^{\infty} b_n n/(\omega^2-n^2) = 0$.   Thus $c_1 = 0$,

$c_2 = b_m/2m^2 - \displaystyle\sum_{n=1, n\neq m}^{\infty} b_n n/m(\omega^2-n^2)$, which when substituted

into the equation for $y(t)$ yields the desired solution.

11. In order to use the results of Problem 10, we must first
     find the Fourier series for the given $f(t)$. Using Eqs.(2)
     and (3) with $\ell = \pi$, we find that
     $a_0 = (1/\pi)\displaystyle\int_0^{\pi} dx - (1/\pi)\displaystyle\int_{\pi}^{2\pi} dx = 0;$

     $a_n = (1/\pi)\displaystyle\int_0^{\pi}\cos nx\,dx - (1/\pi)\displaystyle\int_{\pi}^{2\pi}\cos nx\,dx = 0;$ and

     $b_n = (1/\pi)\displaystyle\int_0^{\pi}\sin nx\,dx - (1/\pi)\displaystyle\int_{\pi}^{2\pi}\sin nx\,dx = 0$ for n even and
     $= 4/n\pi$ for n odd.   Thus

     $f(t) = (4/\pi)\displaystyle\sum_{n=1}^{\infty}\sin(2n-1)t/(2n-1)$.   Comparing this to the

     forcing function of Problem 10 we see that b_n of Problem
     10 has the specific values $b_{2n} = 0$ and
     $b_{2n-1} = (4/\pi)/(2n-1)$ in this example.   Substituting these
     into the answer to Problem 10 yields the desired
     solution.   Note that we have asumed ω is not a positive
     integer.   Note also, that if the solution to Problem 10
     is not available, the procedure for solving Problem 11
     would be exactly the same as shown in Problem 10.

12. In this case the Fourier series for $f(t)$ is given by

     $f(t) = 1/2 + (4/\pi^2)\displaystyle\sum_{n=1}^{\infty}\cos(2n-1)\pi t/(2n-1)^2$ and thus we may

     not use the form of the answer in Problem 10.   The
     procedure outlined there, however, is applicable and will
     yield the desired solution.

14b. Note that f and f' are continuous at all points where f''
     is continuous.   Let $x_1, \ldots, x_m$ be the points in $(-\ell, \ell)$
     where f'' is not continuous.   By splitting up the interval

of integration at these points, and integrating Eq.(3) by parts twice, we obtain

$$n^2 b_n = \frac{n}{\pi} \sum_{i=1}^{m} [f(x_i+) - f(x_i-)] \cos\frac{n\pi x_i}{\ell} - \frac{n}{\pi}[f(\ell-) - f(-\ell+)]\cos n\pi$$

$$-\frac{\ell}{\pi^2} \sum_{i=1}^{m} [f'(x_i+) - f'(x_1-)] \sin\frac{n\pi x_i}{\ell} - \frac{\ell}{\pi^2}\int_{-\ell}^{\ell} f''(x) \sin\frac{n\pi x}{\ell} dx, \quad \text{where}$$

we have used the fact that cosine is continuous. We want the first two terms on the right side to be zero, for otherwise they grow in magnitude with n. Hence f must be continuous throughout the closed interval $[-\ell, \ell]$. The last two terms are bounded, by the hypotheses on f' and f". Hence $n^2 b_n$ is bounded; similarly $n^2 a_n$ is bounded. Convergence of the Fourier series then follows by

comparison with $\sum_{n=1}^{\infty} n^{-2}$.

Section 10.4, Page 540

11. Sketch the graph to assist in answering this problem.

16. Since sine is an odd function and cosine is an even function, Property 3 tells us that the product is an odd function. Hence we use Property 5 to conclude that
$$\int_{-\ell}^{\ell} \sin(n\pi x/\ell) \cos(n\pi x/\ell) dx = 0.$$

All functions and their derivatives in Problems 20 thorugh 30 are piecewise continuous on the given intervals and their extensions. Thus the Fourier Theorem applies in all cases.

20. For the cosine series we use the even extension of the function given in Eq.(13) and hence
$$f(x) = \begin{cases} 0 & -2 \leq x < -1 \\ 1+x & -1 \leq x < 0 \end{cases} \quad \text{on the interval } -2 \leq x < 0.$$
However, we don't really need this, as the coefficients in this case are given by Eqs.(7), which just use the original values for f(x) on $0 < x \leq 2$. Applying Eqs.(7) we have $\ell = 2$ and thus
$$a_0 = (2/2)\int_0^1 (1-x)dx + (2/2)\int_1^2 0 dx = 1/2. \quad \text{Similarly,}$$

$$a_n = (2/2)\int_0^1 (1-x)\cos(n\pi x/2)dx = 4[1-\cos(n\pi/2)]/n^2\pi^2 \text{ and}$$
$$b_n = 0. \quad \text{Substituting these values in the Fourier series}$$

yields the desired results.

For the sine series, we use Eqs.(8) with $\ell = 2$. Thus $a_n = 0$ and

$$b_n = (2/2)\int_0^1 (1-x)\sin(n\pi x/2)dx = 4[n\pi/2 - \sin(n\pi/2)]/n^2\pi^2.$$

21. The graph of the fuction to which the series converges is shown in the figure. Using Eqs.(7) with $\ell = 2$ we have $a_0 = \int_0^1 dx = 1$ and $a_n = \int_0^1 \cos(n\pi x/2)dx = 2\sin(n\pi/2)/n\pi$. Thus $a_n = 0$ for n even, $a_n = 2/n\pi$ for $n = 1,5,9,\ldots$ and $a_n = -2/n\pi$ for $n = 3,7,11,\ldots$. Hence we may write $a_{2n} = 0$ and $a_{2n-1} = 2(-1)^{n+1}/(2n-1)\pi$, which when substituted into the series gives the desired answer.

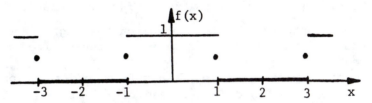

24. The graph of the function to which the series converges is indicated in the figure.

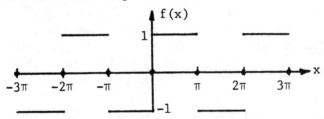

Since we want sine series, we use Eqs.(8) to find, with $\ell = \pi$, that $b_n = (2/\pi)\int_0^\pi \sin nx\,dx = 2[1-(-1)^n]/n\pi$ and thus $b_n = 0$ for n even and $b_n = 4/n\pi$ for n odd.

26. The graph of the function to which the series converges is shown in the figure.

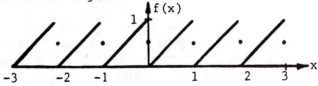

We note that f(x) is specified over its entire fundamental period (T = 1) and hence we cannot extend f to make it either an odd or an even function. Using

Eqs. (2) and (3) from Section 10.3 we have $(l = 1/2)$
$$a_0 = 2\int_0^1 x\,dx = 1, \quad a_n = 2\int_0^1 x\cos(2n\pi x)\,dx = 0 \text{ and}$$

$$b_n = 2\int_0^1 x\sin(2n\pi x)\,dx = -1/n\pi.$$ Substituting these values
into Eq. (1) of Section 10.3 yields the desired solution.
It can also be observed from the above graph that
$g(x) = f(x) - 1/2$ is an odd function. If Eqs. (8) is used
with $g(x)$, then it is found that

$$g(x) = (-1/\pi) \sum_{n=1}^{\infty} \sin(2n\pi x)/n \text{ and thus we obtain the same}$$
series for $f(x)$ as found above.

31. Writing $\int_{-l}^{l} f(x)\,dx = \int_{-l}^{0} f(x)\,dx + \int_{0}^{l} f(x)\,dx$, substituting
$x = -y$ into the first integral on the right, and using
the fact that $f(x) = -f(-x)$ yields the desired result.

32. To prove property 2 let f_1 and f_2 be odd functions and
let $f(x) = f_1(x) \pm f_2(x)$. Then $f(-x) = f_1(-x) \pm f_2(-x) =$
$-f_1(x) \pm [-f_2(x)] = -f_1(x) \mp f_2(x) = -f(x)$, so $f(x)$ is
odd. Now let $g(x) = f_1(x)f_2(x)$, then
$g(-x) = f_1(-x)f_2(-x) = [-f_1(x)][-f_2(x)] = f_1(x)f_2(x) = g(x)$
and thus $g(x)$ is even. Finally, let $h(x) = f_1(x)/f_2(x)$
and hence $h(-x) = f_1(-x)/f_2(-x) = [-f_1(x)]/[-f_2(x)] =$
$f_1(x)/f_2(x) = h(x)$, which says $h(x)$ is also even.
Property 3 is proven in a similar manner.

34. Since $F(x) = \int_0^x f(t)\,dt$ we have
$$F(-x) = \int_0^{-x} f(t)\,dt = -\int_0^x f(-s)\,ds \text{ by letting } t = -s.$$ If f
is an even function, $f(-s) = f(s)$, we then have
$$F(-x) = -\int_0^x f(s)\,ds = -F(x) \text{ from the original definition of}$$
F. Thus $F(x)$ is an odd function. The argument is
similar if f is odd.

35. Set $x = l/2$ in Eq. (6) of Section 10.3. Since we know f
is continuous at the point, we may conclude, by the
Fourier theorem, that the series will converge to
$f(l/2) = l$ at this point. Thus we have

$$l = l/2 + (2l/\pi) \sum_{n=1}^{\infty} (-1)^{n+1}/(2n-1), \text{ since}$$
$$\sin[(2n-1)\pi/2] = (-1)^{n+1}.$$

37a. Multiplying both sides of the equation by $f(x)$ and integrating from 0 to ℓ gives $\int_0^\ell [f(x)]^2 dx =$

$$\int_0^\ell [f(x) \sum_{n=1}^\infty b_n \sin(n\pi x/\ell)] dx = \sum_{n=1}^\infty b_n \int_0^\ell f(x) \sin(n\pi x/\ell) dx =$$

$(\ell/2) \sum_{n=1}^\infty b_n^2$. This result is identical to that of Problem 13 of Section 10.3 if we set $a_n = 0$, $n = 0,1,2,\ldots$. In a similar manner, it can be shown that

$$(2/\ell) \int_0^\ell [f(x)]^2 dx = a_0^2/2 + \sum_{n=1}^\infty a_n^2.$$

37b. Since $f(x) = x$ and $b_n = 2\ell(-1)^{n+1}/n\pi$, we obtain

$$(2/\ell) \int_0^\ell [f(x)]^2 dx = (2/\ell) \int_0^\ell x^2 dx = 2\ell^2/3 = \sum_{n=1}^\infty b_n^2 = \sum_{n=1}^\infty 4\ell^2/n^2\pi^2 =$$

$4\ell^2/\pi^2 \sum_{n=1}^\infty (1/n^2)$ or $\pi^2/6 = \sum_{n=1}^\infty (1/n^2)$.

38. We assume that the extensions of f and f' are piecewise continuous on $[-2\ell, 2\ell]$. Since f is an odd periodic function of fundamental period 4ℓ it follows from properties 2 and 3 that $f(x)\cos(n\pi x/2\ell)$ is odd and $f(x)\sin(n\pi x/2\ell)$ is even. Thus the Fourier coefficients of f are given by Eqs.(8) with ℓ replaced by 2ℓ; that is $a_n = 0$, $n = 0,1,2,\ldots$ and
$b_n = (2/2\ell) \int_0^{2\ell} f(x)\sin(n\pi x/2\ell) dx$, $n = 1,2,\ldots$. The

Fourier sine series for f is $f(x) = \sum_{n=1}^\infty b_n \sin(n\pi x/2\ell)$.

39. From Problem 38 we have $b_n = (1/\ell) \int_0^{2\ell} f(x)\sin(n\pi x/2\ell) dx$

$= (1/\ell) \int_0^\ell f(x)\sin(n\pi x/2\ell) dx + (1/\ell) \int_\ell^{2\ell} f(2\ell-x)\sin(n\pi x/2\ell) dx$

$= (1/\ell) \int_0^\ell f(x)\sin(n\pi x/2\ell) dx + (1/\ell) \int_\ell^0 f(s)\sin[n\pi(2\ell-s)/2\ell] ds$

$= (1/\ell) \int_0^\ell f(x)\sin(n\pi x/2\ell) dx - (1/\ell) \int_0^\ell f(s)\cos(n\pi)\sin(n\pi s/2\ell) ds$

and thus $b_n = 0$ for n even and
$b_n = (2/\ell) \int_0^\ell f(x)\sin(n\pi x/2\ell) dx$ for n odd. The Fourier

series for f is given in Problem 38, where the b_n are given above.

Section 10.5, Page 550

1b. Since the B.C. for this heat conduction problem are $u(0,t) = u(100,t) = 0$, $t > 0$, the solution $u(x,t)$ is given by Eq.(4) with $\alpha^2 = 1.14 \text{cm}^2/\text{sec}$, $\ell = 100$ cm, and the coefficients b_n determined by Eq.(6) with the I.C. $u(x,0) = f(x) = x$, $0 \le x \le 50$; $= 100 - x$, $50 \le x \le 100$.

Thus $b_n = (1/50)[\int_0^{50} x\sin(n\pi x/100)dx +$
$\int_{50}^{100}(100-x)\sin(n\pi x/100)dx] = [400\sin(n\pi/2)]/n^2\pi^2$, $n = 1,2,\ldots$.
It follows that

$$u(x,t) = (400/\pi^2)\sum_{n=1}^{\infty}\sin(n\pi/2)\exp[-n^2\pi^2\alpha^2 t/10000]\sin(n\pi x/100)/n^2.$$

2.   Since the B.C. for this heat conduction problem are $u(0,t) = u(20,t) = 0$, $t > 0$, the solution $u(x,t)$ is given by Eq.(4) with $\ell = 20$ cm, and the coefficients b_n determined by Eq.(6) with the I.C. $u(x,0) = f(x) = 100°\text{C}$. Thus $b_n = (1/10)\int_0^{20}(100)\sin(n\pi x/20)dx = -200[(-1)^n-1]/n\pi$ and hence $b_{2n} = 0$ and $b_{2n-1} = 400/(2n-1)\pi$. Substituting these values into Eq.(4) yields the desired solution. For aluminum, we have $\alpha^2 = .86 \text{ cm}^2/\text{sec}$ (from Table 10.1.1, Sect.10.1) and thus $u(10,30) = (400/\pi)\{\exp[-\pi^2(.86)30/400] - (1/3)\exp[-9\pi^2(.86)30/400]\} = 67.2°\text{C}$.

3b. Using only one term in the series for $u(x,t)$, we must solve the equation $25 = (400/\pi)\exp[-\pi^2(.86)t/400]$ for t. Taking the logarithm of both sides and solving for t yields $t \cong 400\ln(16/\pi)/\pi^2(.86) = 76.7$ sec.

4b. We have from Eq.(9), with $\alpha = 1$
$$u(x,t) = \frac{40}{\pi}[e^{-\pi^2 t}\sin\pi x + \frac{1}{3}e^{-9\pi^2 t}\sin 3\pi x].$$
Setting $x = 1/2$ and $u(1/2,t) = .1$ yields
$\frac{40}{\pi}[e^{-\pi^2 t} - \frac{1}{3}e^{-9\pi^2 t}] = .1$, which is the same as in part a, except for the second term in the brackets. In part a, the correct answer can be found by taking the natural

logarithm of both sides. In this case a numerical
procedure is needed to get the answer precisely.
However, if t ≅ .49108, as in part a, the second term in
the brackets is negligible compared to the first, so the
same value is also correct to five decimal places.

4c. You must recalculate b_n, as given in Eq.(6), with
     f(x) = 10, then substitute in Eq.(4).

5a. Since the B.C. are not homogeneous, we must first find
     the steady state solution. Using Eqs.(12) and (13) we
     have v″ = 0 with v(0) = 0 and v(ℓ) = 60, which has the
     solution v(x) = 60x/ℓ. Thus the transient solution w(x,t)
     satisfies the equations $\alpha^2 w_{xx} = w_t$, w(0,t) = 0,
     w(ℓ,t) = 0 and w(x,0) = 25 - 60x/ℓ, which are obtained
     from Eqs.(16) and (18). The solution of this problem is
     given by Eq.(4) with the b_n given by Eq.(6):
     $$b_n = (2/\ell)\int_0^\ell (25 - 60x)/\ell \, \sin(n\pi x/\ell)\,dx = [50+70(-1)^n]/n\pi.$$
     Finally, u(x,t) = v(x) + w(x,t), which gives the desired
     solution when all the terms are substituted into the
     indicated equations.

5e. A detailed analysis explaining the stated observation is
     given in the paper mentioned in the footnote. The
     analysis is based upon the fact that c_n has distinctly
     different values depending on whether n is even or odd.
     That is $c_n = -20/n\pi$ for n odd and $c_n = 120/n\pi$ for n even
     and thus

$$u(x,t) = 60x/\ell + (60/\pi)\sum_{n=1}^{\infty}(1/n)\exp[-.86(2n)^2\pi^2 t/\ell^2]\sin(n\pi x/\ell)$$

$$- (20/\pi)\sum_{n=1}^{\infty}[1/(2n-1)]\exp[-.86(2n-1)^2\pi^2 t/\ell^2]\sin[(2n-1)\pi x/\ell].$$

It is then shown that the first sum along with the steady
state solution (a temperature redistribution problem)
converges faster than the second sum, which represents a
heat change. The negative sign on this last sum then
indicates that the steady state temperature is approached
from below.

7.   Since the B.C. are $u_x(0,t) = u_x(\ell,t) = 0$, t > 0, the
     solution u(x,t) is given by Eq.(32) with the coefficients
     c_n determined by Eq.(34). Substituting the I.C.
     u(x,0) = f(x) = sin($\pi x/\ell$) into Eq.(34) yields

$$c_0 = (2/l) \int_0^l \sin(\pi x/l) \, dx = 4\pi \text{ and}$$

$$c_n = (2/l) \int_0^l \sin(\pi x/l) \cos(n\pi x/l) \, dx$$

$$= (1/l) \int_0^l \{\sin[(n+1)\pi x/l] - \sin[(n-1)\pi x/l]\} \, dx$$

$$= (1/\pi)\{[1 - \cos(n+1)\pi]/(n+1) - [1 - \cos(n-1)\pi]/(n-1)\}$$

$$= 0, \ n \text{ odd}; \ = -4/(n^2-1)\pi, \ n \text{ even}. \quad \text{Thus}$$

$$u(x,t) = 2/\pi - (4/\pi) \sum_{n=1}^{\infty} \exp[-4n^2\pi^2\alpha^2 t/l^2] \cos(2n\pi x/l)/(4n^2-1)$$

where we are now summing over even terms by setting
n = 2n. As t → ∞ we see that all terms in the series
decay to zero except the constant term, $2/\pi$. Hence
$\lim_{t\to\infty} u(x,t) = 2/\pi$,

9. The steady-state temperature distribution v(x) must
   satisfy Eq.(12) and also satsify the B.C. $v_x(0) = 0$,
   $v(l) = 0$. The general solution of $v'' = 0$ is
   v(x) = Ax + B. The B.C. imply that A = 0 and B = 0 so the
   steady-state solution is v(x) = 0.

11a. Substituting u(x,t) = X(x)T(t) into Eq.(1) leads to the
    two O.D.E. $X'' - \sigma X = 0$ and $T' - \alpha^2\sigma T = 0$. An argument
    similar to the one in the text implies that we must have
    X(0) = 0 and $X'(l) = 0$. Also, by assuming σ is real and
    considering the three cases $\sigma < 0$, $\sigma = 0$, and $\sigma > 0$ we
    can show that only the case $\sigma < 0$ leads to nontrivial
    solutions of $X'' - \sigma X = 0$ with X(0) = 0 and $X'(l) = 0$.
    Setting $\sigma = -\lambda^2$, we obtain $X(x) = k_1\sin\lambda x + k_2\cos\lambda x$.
    Now, X(0) = 0 → k_2 = 0 and thus $X(x) = k_1\sin\lambda x$.
    Differentiating and setting x = l yields $\lambda k_1\cos\lambda l = 0$.
    Since $\lambda = 0$ and $k_1 = 0$ lead to u(x,t) = 0, we must choose
    λ so that $\cos\lambda l = 0$, or $\lambda = (2n-1)\pi/2l$, n = 1,2,3,... .
    These values for λ imply that $\sigma = -(2n-1)^2\pi^2/4l^2$ so the
    solutions T(t) of $T' - \alpha^2\sigma T = 0$ are proportional to
    $\exp[-(2n-1)^2\pi^2\alpha^2 t/4l^2]$. Combining the above results leads
    to the desired set of fundamental solutions.

11b. In order to satisfy the I.C. u(x,0) = f(x) we assume that
    u(x,t) has the form

$$u(x,t) = \sum_{n=1}^{\infty} c_n \exp[-(2n-1)^2\pi^2\alpha^2 t/4l^2] \sin[(2n-1)\pi x/2l]. \quad \text{The}$$

coefficients c_n are determined by the requirement that

$$u(x,0) = \sum_{n=1}^{\infty} c_n \sin[(2n-1)\pi x/2\ell] = f(x).$$ Referring to Problem 39 of Section 10.4 reveals that such a representation for $f(x)$ is possible if we choose the coefficients $c_n = (2/\ell)\int_0^\ell f(x)\sin[(2n-1)\pi x/2\ell]dx$.

12. To reduce the problem to one with homogeneous B.C. we first find the steady state solution by solving $v'' = 0$, subject to the B.C. $v(0) = T$, $v'(\ell) = 0$. We find that $v(x) = T$. As in the text we write $u(x,t) = v(x) + w(x,t)$ and find that w must satisfy the equation $\alpha^2 w_{xx} = w_t$, the B.C. $w(0,t) = w_x(\ell,t) = 0$, and the I.C. $w(x,0) = u(x,0) - v(x) = f(x) - T$. Then a formal series expansion for $w(x,t)$, and hence $u(x,t)$, can be obtained from Problem 11 with $f(x)$ replaced by $f(x) - T$ in determining the Fourier coefficients.

14. We must solve $v_1''(x) = 0$, $0 \le x \le a$ and $v_2''(x) = 0$, $a \le x \le \ell$ subject to the B.C. $v_1(0) = 0$, $v_2(\ell) = T$ and the continuity conditions at $x = a$: $v_1(a) = v_2(a)$ and $-\kappa_1 A_1 v_1'(a) = -\kappa_2 A_2 v_2'(a)$. The general solutions to the two O.D.E. are $v_1(x) = C_1(x) + D_1$ and $v_2(x) = C_2(x) + D_2$. By applying the boundary and continuity conditons we may solve for C_1, D_1, and C_2 and D_2 to obtain the desired solution.

Section 10.6, Page 559

1. Since the initial velocity is zero, the solution is given by Eq.(22) with the coefficients k_n given by Eq.(20). Substituting $f(x)$ into Eq.(20) yields $k_n =$ $(2/\ell)[\int_0^{\ell/2} Ax\sin(n\pi x/\ell)dx + \int_{\ell/2}^\ell A(\ell-x)\sin(n\pi x/\ell)dx] =$ $[4A\ell\sin(n\pi/2)]/n^2\pi^2$. The desired solution is then obtained by substituting these values into Eq.(22).

3. The solution is given by Eq.(18) with the coefficients k_n and c_n givenby Eqs.(20) and (29). Substituting $u(x,0) = f(x) = 0$ into Eq.(20) yields $k_n = 0$, $n = 1,2,\ldots$. From Eq.(29) we see that $c_n = (2/n\pi a)\int_0^\ell g(x)\sin(n\pi x/\ell)dx$, $n = 1,2,\ldots$.

4. Use the results of Problem 3.

6. Assuming that $u(x,t) = X(x)T(t)$ and substituting for u in
    Eq.(1) leads to the pair of O.D.E. $X'' + \sigma X = 0$,
    $T'' + a^2\sigma T = 0$. Applying the B.C. $u(0,t) = 0$ and
    $u_x(l,t) = 0$ to $u(x,t)$ we see that we must have $X(0) = 0$
    and $X'(l) = 0$. By considering the three cases $\sigma < 0$,
    $\sigma = 0$, and $\sigma > 0$ it can be shown that nontrivial
    solutions of the problem $X'' + \sigma X = 0$, $X(0) = 0$, $X'(l) = 0$
    are possible if and only if
    $\sigma = (n-1/2)^2\pi^2/l^2 = (2n-1)^2\pi^2/4l^2$, $n = 1,2,\ldots$ and the
    corresponding solutions for $X(x)$ are proportional to
    $\sin[(2n-1)\pi x/2l]$. Using these values for σ we find that
    $T(t)$ is a linear combination of $\sin[(2n-1)\pi at/2l]$ and
    $\cos[(2n-1)\pi at/2l]$. Now, the I.C. $u_t(x,0)$ implies that
    $T'(0) = 0$ and thus functions of the form
    $u_n(x,t) = \sin[(2n-1)\pi x/2l]\cos[(2n-1)\pi at/2l]$, $n = 1,2,\ldots$
    satsify the P.D.E. (1), the B.C. $u(0,t) = 0$, $u_x(l,t) = 0$,
    and the I.C. $u_t(x,0) = 0$. We now seek a superposition of
    these fundamental solutions u_n that also satisfies the
    I.C. $u(x,0) = f(x)$. Thus we assume that

    $$u(x,t) = \sum_{n=1}^{\infty} c_n\sin[(2n-1)\pi x/2l]\sin[(2n-1)\pi at/2l].$$ The I.C.

    now implies that we must have

    $$f(x) = \sum_{n=1}^{\infty} c_n\sin[(2n-1)\pi x/2l].$$ From Problem 39 of Section

    10.4 we see that $f(x)$ can be represented by such a series
    and that
    $c_n = (2/l)\int_0^l f(x)\sin[(2n-1)\pi x/2l]dx$, $n = 1,2,\ldots$.
    Substituting these values into the above series for
    $u(x,t)$ yields the desired solution.

8. Using the chain rule we obtain $u_x = u_\xi\xi_x + u_\eta\eta_x =$
    $u_\xi + u_\eta$ since $\xi_x = \eta_x = 1$. Differentiating a second time
    gives $u_{xx} = u_{\xi\xi} + 2u_{\xi\eta} + u_{\eta\eta}$. In a similar way we obtain
    $u_t = u_\xi\xi_t + u_\eta\eta_t = -au_\xi + au_\eta$, since $\xi_t = -a$, $\eta_t = a$. Thus
    $u_{tt} = a^2(u_{\xi\xi} - 2u_{\xi\eta} + u_{\eta\eta})$. Substituting for u_{xx} and u_{tt} in
    the wave equation, we obtain $u_{\xi\eta} = 0$. Integrating both
    sides of $u_{\xi\eta} = 0$ with respect to η yields
    $u_\xi(\xi,\eta) = \gamma(\xi)$ where γ is an arbitrary function of ξ.
    Integrating both sides of $u_\xi(\xi,\eta) = \gamma(\xi)$ with respect to ξ
    yields $u(\xi,\eta) = \int\gamma(\xi)d\xi + \psi(\eta) = \phi(\xi) + \psi(\eta)$ where $\psi(\eta)$

is an arbitrary function of η and $\int\gamma(\xi)\,d\xi$ is some
function of ξ denoted by $\phi(\xi)$.   Thus
$u(x,t) = u(\xi(x,t),\eta(x,t)) = \phi(x - at) + \psi(x + at)$.

9.   The graph of $y = \sin(x-at)$ for the various values of t is
     indicated in the figure below. Note that the graph of
     $y = \sin x$ is displaced to the right by the distance "at"
     for each value of t.

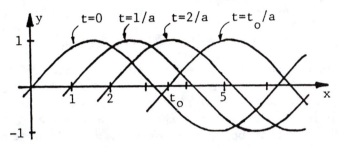

     Similarly, the graph of $y = \phi(x + at)$ would be displaced
     to the left by a distance "at" for each t.   Thus
     $\phi(x + at)$ represents a wave moving to the left.

11. Write the equation as $a^2 u_{xx} = u_{tt} + \alpha^2 u$ and assume
     $u(x,t) = X(x)T(t)$. The separation constant is $-\lambda^2$ using
     the same arguments as in the text and earlier problems.

12b. Using the hint and the first equation obtained in part
     (a) leads to $\phi(x) + \psi(x) = 2\phi(x) + c = f(x)$ so
     $\phi(x) = (1/2)f(x) - c/2$ and $\psi(x) = (1/2)f(x) + c/2$. Hence
     $u(x,t) = \phi(x - at) + \psi(x + at) = (1/2)[f(x - at) - c] +$
     $(1/2)[f(x + at) + c] = (1/2)[f(x - at) + f(x + at)]$.

12c. Substituting $x + at$ for x in $f(x)$ yields
     $$f(x + at) = \begin{cases} 2 & -1 < x + at < 1 \\ 0 & \text{otherwise} \end{cases}.$$
     Subtracting "at" from both sides of the inequality then
     yields
     $$f(x + at) = \begin{cases} 2 & -1 - at < x < 1 - at \\ 0 & \text{otherwise} \end{cases}.$$

13b. From part (a) we have $\psi(x) = -\phi(x)$ and hence
     $$\psi(x) = (1/2a)\int_{x_0}^{x} g(\xi)\,d\xi - \phi(x_0).$$

13c. $u(x,t) = \phi(x-at) + \psi(x+at)$

$\quad = -(1/2a)\int_{x_0}^{x-at} g(\xi)\,d\xi + \phi(x_0) + (1/2a)\int_{x_0}^{x+at} g(\xi)\,d\xi - \phi(x_0)$

$\quad = (1/2a)[\int_{x_0}^{x+at} g(\xi)\,d\xi - \int_{x_0}^{x-at} g(\xi)\,d\xi]$

$\quad = (1/2a)[\int_{x_0}^{x+at} g(\xi)\,d\xi + \int_{x-at}^{x_0} g(\xi)\,d\xi]$

$\quad = (1/2a\int_{x-at}^{x+at} g(\xi)\,d\xi.$

14. Use the result of Problem 5.

17. Substituting $u(r,\theta,t) = R(r)\Theta(\theta)T(t)$ into the P.D.E.
     yields $R''\Theta T + R'\Theta T/r + R\Theta''T/r^2 = R\Theta T''/a^2$ or equivalently
     $R''/R + R'/rR + \Theta''/r^2 = T''/a^2T$. In order for this equation
     to be valid for $0 < r < r_0$, $0 \leq \theta \leq 2\pi$, $t > 0$, it is
     necessary that both sides of the equation be equal to the
     same constant $-\sigma$. Otherwise, by keeping r and θ fixed
     and varying t, one side would remain unchanged while the
     other side varied. Thus we arrive at the two equations
     $T'' + \sigma a^2 T = 0$ and $r^2R''/R + rR'/R + \sigma r^2 = -\Theta''/\Theta$. By an
     argument similar to the one above we conclude that both
     sides of the last equation must be equal to the same
     constant δ. This leads to the two equations
     $r^2R'' + rR' + (\sigma r^2 - \delta)R = 0$ and $\Theta'' + \delta\Theta = 0$. Since the
     circular membrane is continuous, we must have
     $\Theta(2\pi) = \Theta(0)$, which requires $\delta = \mu^2$, μ a non-negative
     integer. The condition $\Theta(2\pi) = \Theta(0)$ is also known as
     the periodicity condition. Since we also desire
     solutions which vary periodically in time, it is clear
     that the separation constant σ should be positive,
     $\sigma = \lambda^2$. Thus we arrive at the three equations
     $r^2R'' + rR' + (\lambda^2 r^2 - \mu^2)R = 0$, $\Theta'' + \mu^2\Theta = 0$, and
     $T'' + \lambda^2 a^2 T = 0$.

Section 10.7, Page 569

1a. Assuming that $u(x,y) = X(x)Y(y)$ leads to the two O.D.E.
     $X'' - \sigma X = 0$, $Y'' + \sigma Y = 0$. The B.C. $u(0,y) = 0$,
     $u(a,y) = 0$ imply that $X(0) = 0$ and $X(a) = 0$. Thus
     nontrivial solutions to $X'' - \sigma X = 0$ which satisfy these
     boundary conditions are possible only if $\sigma = -(n\pi/a)^2$,
     $n = 1,2\ldots$; the corresponding solutions for $X(x)$ are
     proportional to $\sin(n\pi x/a)$. The B.C. $u(x,0) = 0$ implies
     that $Y(0) = 0$. Solving $Y'' - (n\pi/a)^2 Y = 0$ subject to this
     condition we find that $Y(y)$ must be proportional to

$\sinh(n\pi y/a)$. The fundamental solutions
$u_n(x,y) = \sin(n\pi x/a)\sinh(n\pi y/a)$, $n = 1,2,\ldots$ satisfy
Laplace's equation and the homogeneous B.C. We assume

that $u(x,y) = \displaystyle\sum_{n=1}^{\infty} c_n \sin(n\pi x/a)\sinh(n\pi y/a)$, where the

coefficients c_n are determined from the B.C.

$u(x,b) = g(x) = \displaystyle\sum_{n=1}^{\infty} c_n \sin(n\pi x/a)\sinh(n\pi b/a)$. It follows

that

$c_n \sinh(n\pi b/a) = (2/a)\displaystyle\int_0^a g(x)\sin(n\pi x/a)\,dx$, $n = 1,2,\ldots$.

1b. Substituting for $g(x)$ in the equation for c_n we have

$c_n \sinh(n\pi b/a) = (2/a)\,[\displaystyle\int_0^{a/2} x\sin(n\pi x/a)\,dx +$

$\displaystyle\int_{a/2}^a (a-x)\sin(n\pi x/a)\,dx] = [4a\,\sin(n\pi/2)]/n^2\pi^2$, $n = 1,2,\ldots,$

so $c_n = [4a\,\sin(n\pi/2)]/[n^2\pi^2\sinh(n\pi b/a)]$. Substituting
these values for c_n in the above series yields the
desired solution.

2. In solving the D.E. $Y'' - \lambda^2 Y = 0$, one normally writes
$Y(y) = c_1\sinh\lambda y + c_2\cosh\lambda y$. However, since we have
$Y(b) = 0$, it is advantageous to rewrite Y as
$Y(y) = d_1\sinh\lambda(b-y) + d_2\cosh\lambda(b-y)$, where d_1, d_2 are also
arbitrary constants and can be related to c_1, c_2 using
the appropriate hyperbolic trigonometric identities. The
important thing, however, is that the second form also
satisfies the D.E. and thus $Y(y) = d_1\sinh\lambda(b-y)$ satisfies
the D.E. and the homogeneous B.C. $Y(b) = 0$. The rest of
the problem follows the pattern of Problem 1.

4. Following the pattern of Problem 3, one would consider
adding the solutions of four problems, each with only one
non-homogeneous B.C. It is also possible to consider
adding the solutions of only two problems, each with only
two non-homogeneous B.C., as long as they involve the
same variable. For instance, one such problem would be
$u_{xx} + u_{yy} = 0$, $u(x,0) = 0$, $u(x,b) = 0$, $u(0,y) = k(y)$,
$u(a,y) = f(y)$, which has the fundamental solutions
$u_n(x,y) = [c_n\sinh(n\pi x/b) + d_n\cosh(n\pi x/b)]\sin(n\pi y/b)$.

Assuming $u(x,y) = \displaystyle\sum_{n=1}^{\infty} u_n(x,y)$ and using the B.C.

$u(0,y) = k(y)$ we obtain $d_n = (2/a)\int_0^a k(y)\sin(n\pi y/b)\,dy$.

Using the B.C. $u(a,y) = f(y)$ we obtain

$c_n\sinh(n\pi a/b) + d_n\cosh(n\pi a/b) = (2/a)\int_0^a f(y)\sin(n\pi y/b)\,dy$,

which can be solved for c_n, since d_n is already known.

5. Using Eq.(17) and following the same arguments as
    presented in the text, we find that $R(r) = k_1 r^n + k_2 r^{-n}$
    and $\Theta(\theta) = c_1\sin n\theta + c_2\cos n\theta$, for n a positive integer,
    and $u_0(r,\theta) = 1$ for $n = 0$. Since we require that $u(r,\theta)$
    be bounded as $r \to \infty$, we conclude that $k_1 = 0$. The
    fundamental solutions are therefore
    $u_n(r,\theta) = r^{-n}\cos n\theta$, $v_n(r,\theta) = r^{-n}\sin n\theta$, $n = 1,2,\ldots$
    together with $u_0(r,\theta) = 1$. Assuming that u can be
    expressed as a linear combination of the fundamental
    solutions we have

    $u(r,\theta) = c_0/2 + \displaystyle\sum_{n=1}^{\infty} r^{-n}(c_n\cos n\theta + k_n\sin n\theta)$. The B.C.

    requires that

    $u(a,\theta) = c_0/2 + \displaystyle\sum_{n=1}^{\infty} a^{-n}(c_n\cos n\theta + k_n\sin n\theta) = f(\theta)$ for

    $0 \le \theta < 2\pi$. This is precisely the Fourier series
    representation for $f(\theta)$ of period 2π and thus
    $a^{-n}c_n = (1/\pi)\int_0^{2} f(\theta)\cos n\theta\,d\theta$, $n = 0,1,2,\ldots$ and

    $a^{-n}k_n = (1/\pi)\int_0^{2} f(\theta)\sin n\theta\,d\theta$, $n = 1,2\ldots$.

7. Again we let $u(r,\theta) = R(r)\Theta(\theta)$ and thus we have
    $r^2 R'' + rR' - \sigma R = 0$ and $\Theta'' + \sigma\Theta = 0$, with $R(0)$ bounded and
    the B.C. $\Theta(0) = \Theta(\alpha) = 0$. Since $u(r,\theta)$ must be single
    valued we conclude that $\sigma = \lambda^2$ (λ^2 real) and thus
    $\Theta(\theta) = c_1\cos\lambda\theta + c_2\sin\lambda\theta$. The B.C. $\Theta(0) = 0 \to c_1 = 0$ and
    the B.C. $\Theta(\alpha) = 0 \to \lambda = n\pi/\alpha$, $n = 1,2,\ldots$.
    Substituting these values into Eq.(28) we obtain
    $R(r) = k_1 r^{n\pi/\alpha} + k_2 r^{-n\pi/\alpha}$. However $k_2 = 0$ since $R(0)$ must
    be bounded, and thus the fundamental solutions are
    $u_n(r,\theta) = r^{n\pi/\alpha}\sin(n\pi\theta/\alpha)$. The desired solution may now
    be formed using previously discussed procedures.

8. Since neither sinhy nor coshy are bounded as $y \to \infty$, we
    must write the solution to $Y'' - (n\pi/a)^2 Y = 0$ as
    $Y(y) = c_1\exp[n\pi y/a] + c_2\exp[-n\pi y/a]$. Thus we must choose
    $c_1 = 0$ so that $u(x,y) = X(x)Y(y) \to 0$ as $y \to \infty$.

13. Assuming that $u(x,y) = X(x)Y(y)$ and substituting into Eq.(1) leads to the two O.D.E. $X'' - \sigma X = 0$, $Y'' + \sigma Y = 0$. The B.C. $u(x,0) = 0$, $u_y(x,b) = 0$ imply that $Y(0) = 0$ and $Y'(b) = 0$. For nontrivial solutions to exist for $Y'' + \sigma Y = 0$ with these B.C. we find that σ must take the values $(2n-1)^2\pi^2/b^2$, $n = 1,2,...$; the corresponding solutions for $Y(y)$ are proportional to $\sin[(2n-1)\pi y/b]$. Solutions to $X'' - [(2n-1)^2\pi^2/b^2]X = 0$ are of the form $X(x) = A\sinh[(2n-1)\pi x/2b] + B\cosh[(2n-1)\pi x/2b]$. However, the boundary condition $u(0,y) = 0$ implies that $X(0) = B = 0$. It follows that the fundamental solutions are $u_n(x,y) = c_n\sinh[(2n-1)\pi x/2b]\sin[(2n-1)\pi y/2b]$, $n = 1,2,...$. To satisfy the remaining B.C. at $x = a$ we assume that we can represent the solution $u(x,y)$ in the form $u(x,y) = \sum_{n=1}^{\infty}c_n\sinh[(2n-1)\pi x/2b]\sin[(2n-1)\pi y/2b]$.

The coefficients c_n are determined by the B.C.

$$u(a,y) = \sum_{n=1}^{\infty}c_n\sinh[(2n-1)\pi a/2b]\sin[(2n-1)\pi y/2b] = f(y).$$

By properly extending f as a periodic function of period $4b$ as in Problem 39, Section 10.4, we find that the coefficients c_n are given by

$$c_n\sinh[(2n-1)\pi a/2b] = (2/b)\int_0^b f(y)\sin[(2n-1)\pi y/2b]\,dy,$$

$n = 1,2,...$.

192

CHAPTER 11

<u>Section 11.1, Page 582</u>

2. Since the B.C. at $x = 1$ is nonhomogeneous, the B.V.P. is nonhomogeneous.

4. The D.E. may be written $y'' + (\lambda - x^2)y = 0$ and is thus homogeneous, as are both B.C.

5. Since the D.E. contains the nonhomogeneous term 1, the B.V.P. is nonhomogeneous.

7b. Eq.(iii) is linear and separable. Using the latter approach we have $d\mu/\mu = [(Q-P')/P]dx$ and thus $\ln\mu = \int_{x_0}^{x} [Q(s)/P(s)]ds - \ln P$. Taking the exponential of both sides yields Eq.(iv). The choice of x_0 simply alters the constant of integration, which is immaterial here.

9. Since $P(x) = x^2$ and $Q(x) = x$, we find that $\mu(x) = (1/x^2)\exp[\int_{x_0}^{x}(s/s^2)ds] = k/x$, where k is an arbitrary constant which may be set equal to 1. It follows that Bessel's equation takes the form $(xy')' + (x-\upsilon^2/x)y = 0$.

<u>Section 11.2, Page 589</u>

1. If $\lambda < 0$, the general solution of the D.E. is $y = c_1\sinh\sqrt{\mu}\,x + c_2\cosh\sqrt{\mu}\,x$ where $-\lambda = \mu$. The two B.C. require that $c_2 = 0$ and $c_1 = 0$ so $\lambda < 0$ is not an eigenvalue. If $\lambda = 0$, the general solution of the D.E. is $y = c_1 + c_2x$. The B.C. require that $c_1 = 0$, $c_2 = 0$ so $\lambda = 0$ is not an eigenvalue. If $\lambda > 0$, the general solution of the D.E. is $y = c_1\sin\sqrt{\lambda}\,x + c_2\cos\sqrt{\lambda}\,x$. The B.C. require that $c_2 = 0$, $c_1\cos\sqrt{\lambda} = 0$. The second condition is satisfied for $c_1 \neq 0$ if $\lambda = [(2n-1)\pi/2]^2$, $n = 1,2,\ldots$. Thus the eigenvalues are $\lambda_n = [(2n-1)\pi/2]^2$, $n = 1,2,\ldots$ with corresponding eigenfunctions $\phi_n(x) = \sin[(2n-1)\pi x/2]$, $n = 1,2,\ldots$.

4. Note that the problem is identical to Problem 1 with λ replaced by $-\lambda$.

5. If $\lambda = 0$, the general solution of the D.E. is $y = c_1 + c_2 x$. The B.C. $y(0) + y'(0) = 0$ requires $c_1 + c_2 = 0$ and the B.C. $y(1) = 0$ requires $c_1 + c_2 = 0$ and thus $\lambda = 0$ is an eigenvalue with corresponding eigenfunction $\phi_0(x) = 1-x$.

If $\lambda < 0$, set $-\lambda = \mu^2$ to obtain $y = c_1 \cos\mu x + c_2 \sin\mu x$. In this case the B.C. require $c_1 + \mu c_2 = 0$ and $c_1 \cos\mu + c_2 \sin\mu = 0$ which yields nontrivial solutions for c_1 and c_2 (i.e., $c_1 = -\mu c_2$) if and only if $\tan\mu = \mu$. By plotting on the same graph $f(\mu) = \mu$ and $g(\mu) = \tan\mu$, we see that they intersect at $\mu_0 = 0$ ($\mu = 0 \to \lambda = 0$, which has already been discussed), $\mu_1 \cong 4.493$ (which is just to the left of the vertical asymptote of $\tan\mu$ at $\mu = (3\pi/2)$, and for larger values $\mu_n \cong (2n+1)\pi/2$. Since $\lambda_n = -\mu_n^2$, we have $\lambda_1 \cong -20.2$, $\lambda_n \cong -(2n+1)^2\pi^2/4$ and $\phi_n = \sin\mu_n x - \mu_n \cos\mu_n x$.

If $\lambda > 0$, the general solution of the D.E. is $y(x) = c_1 \cosh\sqrt{\lambda}\, x + c_2 \sinh\sqrt{\lambda}\, x$. The B.C. respectively require that $c_1 + \sqrt{\lambda}\, c_2 = 0$ and $c_1 \cosh\sqrt{\lambda} + c_2 \sinh\sqrt{\lambda} = 0$ and thus λ must satisfy $\tanh\sqrt{\lambda} = \sqrt{\lambda}$ in order to have nontrivial solutions. The only solution of this equation is $\lambda = 0$ and thus there are no positive eigenvalues.

8. If $\lambda > 0$, the general solution of the D.E. is $y = c_1 \sin\sqrt{\lambda}\, x + c_2 \cos\sqrt{\lambda}\, x$. The B.C. require that $c_2 - \sqrt{\lambda}\, c_1 = 0$, $(\sin\sqrt{\lambda} + \sqrt{\lambda} \cos\sqrt{\lambda})c_1 + (\cos\sqrt{\lambda} - \sqrt{\lambda} \sin\sqrt{\lambda})c_2 = 0$. In order to have nontrivial solutions λ must satisfy $(\lambda-1)\sin\sqrt{\lambda} - 2\sqrt{\lambda} \cos\sqrt{\lambda} = 0$. In this case $c_2 = \sqrt{\lambda}\, c_1$ and thus $\phi_n = \sin\sqrt{\lambda_n}\, x + \sqrt{\lambda_n} \cos\sqrt{\lambda_n}\, x$. If $\lambda \neq 1$, the eigenvalue equation is equivalent to $\tan\sqrt{\lambda} = 2\sqrt{\lambda}/(\lambda-1)$ and thus by graphing $f(\sqrt{\lambda}) = \tan\sqrt{\lambda}$ and $g(\sqrt{\lambda}) = 2\sqrt{\lambda}/(\lambda-1)$ we can estimate the eigenvalues. Since $g(\sqrt{\lambda})$ has a vertical asymptote at $\lambda = 1$ and $f(\sqrt{\lambda})$ has a vertical asymptote at $\sqrt{\lambda} = \pi/2$, we see that $1 < \sqrt{\lambda_1} < \pi/2$. By interating numerically, we find

$\sqrt{\lambda_1} \cong 1.3$ and thus $\lambda_1 \cong 1.69$. For large values of n, we see from the graph that $\sqrt{\lambda_n} \cong (n-1)\pi$, which are the zeros of $\tan\sqrt{\lambda}$. Thus $\lambda_n \cong (n-1)^2\pi^2$ for large n. For $\lambda \leq 0$, the discussion follows the pattern of earlier problems.

10a. Assuming $y = s(x)u$, we have $y' = s'u + su'$ and $y'' = s''u + 2s'u' + su''$ and thus the D.E. becomes $su'' + (2s'+4s)u' + [s'' + 4s' + (4+9\lambda)s]u = 0$. Setting $2s' + 4s = 0$ we find $s(x) = e^{-2x}$ and the D.E. becomes $u'' + 9\lambda u = 0$. The B.C. $y(0) = 0$ yields $s(0)u(0) = 09$, or $u(0) = 0$ since $s(0) \neq 0$. The B.C. at l is $y'(l) = s'(l)u(l) + s(l)u'(l) = e^{-2l}(-2u(l) + u'(l)) = 0$ and thus $u'(l) - 2u(l) = 0$. Thus the B.V.P. satisfied by $u(x)$ is $u'' + 9\lambda u = 0$, $u(0) = 0$, $u'(l) - 2u(l) = 0$.

If $\lambda < 0$, the general solution of the D.E. $u'' + 9\lambda u = 0$ is $u = c_1\sinh3\mu x + c_2\cosh3\mu x$ where $-\lambda = \mu^2$. The B.C. require that $c_2 = 0$, $c_1(3\mu\cosh3\mu l - 2\sinh3\mu l) = 0$. In order to have nontrivial solutions μ must satisfy the equation $3\mu/2 = \tanh3\mu l$. A graphical analysis reveals that for $l \leq 1/2$ this equation has no solutions for $\mu \neq 0$ so there are no negative eigenvalues for $l \leq 1/2$. If $l > 1/2$ there is one solution and hence one negative eigenvalue with eigenfuction $\phi_{-1}(x) = e^{-2x}\sinh3\mu x$.

If $\lambda = 0$, the general solution of the D.E. $u'' + 9\lambda u = 0$ is $u = c_1 + c_2x$. The B.C. require that $c_1 = 0$, $c_2(1-2l) = 0$ so nontrivial solutions are possible only if $l = 1/2$. In this case the eigenfuction is $\phi_0(x) = xe^{-2x}$.

If $\lambda > 0$, the general solution of the D.E. $u'' + 9\lambda u = 0$ is $u = c_1\sin3\sqrt{\lambda}\,x + c_2\cos\sqrt{\lambda}\,x$. The B.C. require that $c_2 = 0$, $c_1(3\sqrt{\lambda}\cos3\sqrt{\lambda}\,l - 2\sin3\sqrt{\lambda}\,l) = 0$. In order to have nontrivial solutions λ must satisfy the equation $\sqrt{\lambda} = (2/3)\tan3\sqrt{\lambda}\,l$. A graphical analysis reveals that there is an infinite number of solutions to this eigenvalue equation. Thus the eigenfunctions are $\phi_n(x) = e^{-2x}\sin3\sqrt{\lambda_n}\,x$ where the eigenvalues λ_n satisfy $\sqrt{\lambda_n} = (2/3)\tan3\sqrt{\lambda_n}\,l$.

12. This is an Euler equation. If $\lambda = 1$ the general solution of the D.E. is $y = c_1 x + c_2 x \ln x$ and the B.C. require that $c_1 = c_2 = 0$ and thus $\lambda = 1$ is not an eigenvalue. If $\lambda \neq 1$, $y = c_1 x + c_2 x^\lambda$ is the general solution and the B.C. require that $c_1 + c_2 = 0$ and $2c_1 + c_2 2^\lambda - (c_1 + \lambda c_2 2^{\lambda-1}) = 0$. Thus nontrivial solutions exist if and only if $\lambda = 2(1-2^{-\lambda})$. The graphs of $f(\lambda) = \lambda$ and $g(\lambda) = 2(1-2^{-\lambda})$ intersect only at $\lambda = 1$ (which has already been discussed) and $\lambda = 0$. Thus the only eigenvalue is $\lambda = 0$ with corresponding eigenfunction $\phi(x) = x - 1$ (since $c_1 = -c_2$).

14a. For positive λ, the general solution of the D.E. is $y = c_1 \sin\sqrt{\lambda}\, x + c_2 \cos\sqrt{\lambda}\, x$. The B.C. require that $\sqrt{\lambda}\, c_1 + \alpha c_2 = 0$, $c_1 \sin\sqrt{\lambda} + c_2 \cos\sqrt{\lambda} = 0$. Nontrivial solutions exist if and only if $\sqrt{\lambda}\cos\sqrt{\lambda} - \alpha\sin\sqrt{\lambda} = 0$. If $\alpha = 0$ this equation is satisfied by the sequence $\lambda_n = [(2n-1)\pi/2]^2$, $n = 1,2,\ldots$. If $\alpha \neq 0$, λ must satisfy the equation $\sqrt{\lambda}/\alpha = \tan\sqrt{\lambda}$. A plot of the graphs of $f(\sqrt{\lambda}) = \sqrt{\lambda}/\alpha$ and $g(\sqrt{\lambda}) = \tan\sqrt{\lambda}$ reveals that there is an infinite sequence of postive eigenvalues for $\alpha < 0$ and $\alpha > 0$.

14b. By procedures shown previously, the cases $\lambda < 0$ and $\lambda = 0$, when $\alpha < 1$, lead to only the trivial solution and thus by part a all real eigenvalues are positive. For $0 < \alpha < 1$, the graphs of $f(\sqrt{\lambda})$ and $g(\sqrt{\lambda})$ (see part a) intersect once on $0 < \sqrt{\lambda} < \pi/2$. As α approaches 1 from below, the slope of $f(\sqrt{\lambda})$ decreases and thus the intersection point approaches zero.

15. Using the D.E. for ϕ_m and following the hint yields:
$\int_0^l \phi_m'' \phi_n\, dx + \lambda_m \int_0^l \phi_m \phi_n\, dx = 0$. Integrating the first term by parts yields: $\phi_m' \phi_n \Big|_0^l - \int_0^l \phi_m' \phi_n'\, dx = -\lambda_m \int_0^l \phi_m \phi_n\, dx$. Upon utilizing the B.C. the first term on the left vanishes and thus $\int_0^l \phi_n' \phi_m'\, dx = \lambda_m \int_0^l \phi_m \phi_n\, dx$. Similarly, the D.E. for ϕ_n yields $\int_0^l \phi_m' \phi_n'\, dx = \lambda_n \int_0^l \phi_n \phi_m\, dx$ and thus $(\lambda_n - \lambda_m) \int_0^l \phi_m \phi_n\, dx = 0$. If $\lambda_n \neq \lambda_m$ the desired result follows.

16b. The general solution of the D.E. is

$y = c_1 \sin\mu x + c_2 \cos\mu x + c_3 \sinh\mu x + c_4 \cosh\mu x$ where $\lambda = \mu^4$.

The B.C. require that $c_2 + c_4 = 0$, $-c_2 + c_4 = 0$,

$c_1 \sin\mu l + c_2 \cos\mu l + c_3 \sinh\mu l + c_4 \cosh\mu l = 0$, and

$c_1 \cos\mu l - c_2 \sin\mu l + c_3 \cosh\mu l + c_4 \sinh\mu l = 0$. The first two
equations yield $c_2 = c_4 = 0$, and the last two have
nontrivial solutions if and only if
$\sin\mu l \cosh\mu l - \cos\mu l \sinh\mu l = 0$. In this case the third
equation yields $c_3 = -c_1 \sin\mu l/\sinh\mu l$ and thus the desired
eigenfunctions are obtained. The quantity μl can be
approximated by finding the intersection of $f(x) = \tan x$
and $g(x) = \tanh x$, where $x = \mu l$.

Section 11.3, Page 600

1.   In Problem 1 of Section 11.2 we found the eigenfuctions
to be $\sin[(2n-1)\pi x/2]$, $n = 1, 2, \ldots$ and thus we must
choose k_n so that $\int_0^1 \{k_n \sin[(2n-1)\pi x/2]\}^2 dx = 1$, since the
weight function $r(x) = 1$ [See Eq. (20)]. Evaluating the
integral yields $k_n^2/2 = 1$ and thus $k_n = \sqrt{2}$ and the
desired normalized eigenfuctions are obtained.

3.   Note here that $\phi_0(x) = 1$ satisifes Eq. (34) and hence it
is already normalized.

5.   From Problem 9 of Section 11.2 we have $e^x \sin n\pi x$,
$n = 1, 2, \ldots$ as the eigenfunctions and thus k_n must be
chosen so that $\int_0^1 k_n^2 e^{2x} \sin^2 n\pi x \, dx = 1$ [again $r(x) = 1$].
Setting $\sin^2 n\pi x = (1 - \cos 2n\pi x)/2$ and integrating by
parts yields $k_n^2 (e^2-1) n^2\pi^2/4(n^2\pi^2+1) = 1$. Solving for k_n
and multiplying the above eigenfuctions yields the
desired normalized eigenfunctions.

7.   Using Eq. (34) with $r(x) = 1$, we find that the
coefficients of the series (32) are determined by

$a_n = (f, \phi_n) = \sqrt{2} \int_0^1 x \sin[(2n-1)\pi x/2] dx$

$= (4\sqrt{2}/(2n-1)^2\pi^2) \sin(2n-1)\pi/2$. Thus Eq. (32) yields

$$f(x) = \frac{4\sqrt{2}}{\pi^2} \sum_{n=1}^{\infty} \frac{(-1)^{n-1}}{(2n-1)^2} \sqrt{2} \sin[(2n-1)\pi x/2], \quad 0 \le x \le 1,$$

which agrees with the expansion using the approach
developed in Problem 39 of Section 10.4.

10. In this case $\phi_n(x) = (\sqrt{2}/\alpha_n)\cos\sqrt{\lambda_n}\,x$, where
    $\alpha_n = (1 + \sin^2\sqrt{\lambda_n})^{1/2}$. Thus Eq.(34) yields
    $a_n = (\sqrt{2}/\alpha_n)\int_0^1 \cos\sqrt{\lambda_n}\,x\,dx = \sqrt{2}\,\sin\sqrt{\lambda_n}\,/\alpha_n\sqrt{\lambda_n}$.

14. In this case $L[y] = y'' + y' + 2y$ is not of the form shown
    in Eq.(3) and thus the B.V.P. is not self adjoint.

17. In this case $L[y] = [(1+x^2)y']' + y$ and thus the D.E. has
    the form shown in Eq.(3). However, the B.C. are not
    separated and thus we must determine by integration
    whether Eq.(8) is satisfied. Therefore, for u and v
    satisfying the B.C., integration by parts yields the
    following:

$$(L[u],v) = \int_0^1 \{[(1+x^2)u']'+u\}v\,dx = vu'(1+x^2)\Big|_0^1 - \int_0^1 \{(1+x^2)v'u'+uv\}dx$$

$$= vu'(1+x^2)\Big|_0^1 - uv'(1+x^2)\Big|_0^1 + \int_0^1 \{[(1+x^2)v']'+v\}u\,dx$$

$$= (u,L[v])$$

since the integrated terms add to zero with the given
B.C. Thus the B.V.P. is self-adjoint.

21a. Substituting ϕ for y in the D.E., multiplying both sides
     by ϕ, and integrating form 0 to 1 yields

$$\lambda\int_0^1 r\phi^2 dx = \int_0^1 \{-[p(x)\phi']'\phi + q(x)\phi^2\}dx. \quad \text{Integrating the}$$

first term on the right side once by parts, we obtain

$$\lambda\int_0^1 r\phi^2 dx = -p(1)\phi'(1)\phi(1) + p(0)\phi'(0)\phi(0) + \int_0^1 (p\phi'^2+q\phi^2)\,dx.$$

If $a_2 \neq 0$, $b_2 \neq 0$, then $\phi'(1) = -b_1\phi(1)/b_2$ and
$\phi'(0) = -a_1\phi(0)/a_2$ and the result follows. If $a_2 = 0$,
then $\phi(0) = 0$ and the boundary term at 0 will be missing.
A similar result is obtained if $b_2 = 0$.

23a. Using $\phi(x) = u(x) + iv(x)$ in Eq.(4) we have
     $L[\phi] = L[u(x) + iv(x)] = \lambda r(x)[u(x)+iv(x)]$. Using the
     linearity of L and the fact that λ and $r(x)$ are real we
     have $L[u(x)] + iL[v(x)] = \lambda r(x)u(x) + i\lambda r(x)v(x)$.
     Equating the real and imaginary parts shows that both u
     and v satisfy Eq.(1). The B.C. Eq.(2) are also satisfied
     by both u and v, using the same arguments, and thus both

u and v are eigenfunctions.

23c. From part b we have $v(x) = cu(x)$ and thus
$\phi(x) = u(x) + icu(x) = (1+ic)u(x)$.

24b. See Problem 9 of Section 11.2.

24c. The eigenfuctions are $\phi_n(x) = e^x\sin n\pi x$, $n = 1,2...$ and
the weight function is $r(x) = 1$. Thus
$\int_0^1 r(x)\phi_n(x)\phi_m(x)\,dx = \int_0^1 e^{2x}\sin(n\pi x)\sin(m\pi x)\,dx$, which, by
direct integration, is not necessarily zero when $m \neq n$.

25. If $\lambda = 1$, the general solution to the D.E. is
$y = c_1 x + c_2 x \ln x$. The B.C. require that
$c_1 = 0$, $2c_1 + 2(\ln 2)c_2 = 0$ so $c_1 = c_2 = 0$ and $\lambda = 1$ is
not an eigenvalue. If $\lambda \neq 1$, the general solution to the
D.E. is $y = c_1 x + c_2 x^\lambda$. The B.C. require that $c_1 + c_2 = 0$,
$2c_1 + 2^\lambda c_2 = 0$. Nontrivial solutions exist if and only
if $2^\lambda - 2 = 0$. If λ is real, this equation has no
solution (other than $\lambda = 1$) and again $y = 0$ is the only
solution to the boundary value problem. Suppose that
$\lambda = a + bi$ with $b \neq 0$. Then
$2^\lambda = 2^{a+bi} = 2^a 2^{bi} = 2^a \exp(ib\ln 2)$, which upon substitution
into $2^\lambda = 2$ yields the equation $\exp(ib\ln 2) = 2^{1-a}$. It
follows that $a = 1$ and $b(\ln 2) = 2n\pi$ or $b = 2n\pi/\ln 2$,
$n = \pm 1, \pm 2, \ldots$. Thus the only eigenvalues of the problem
are $\lambda_n = 1 + i(2n\pi/\ln 2)$, $n = \pm 1, \pm 2, \ldots$.

26b. The general solution of the D.E. is
$y = c_1 + c_2 x + c_3 \sin\sqrt{\lambda}\, x + c_4 \cos\sqrt{\lambda}\, x$. The B.C. require
that
$c_1 + c_4 = 0$, $c_4 = 0$, $c_1 + c_2\ell + c_3\sin\sqrt{\lambda}\,\ell + c_4\cos\sqrt{\lambda}\,\ell = 0$, and
$c_2 + \sqrt{\lambda}\,c_3\cos\sqrt{\lambda}\,\ell - \sqrt{\lambda}\,c_4\sin\sqrt{\lambda}\,\ell = 0$. Thus $c_1 = c_4 = 0$
and for nontrivial solutions to exist λ must satisfy the
equation $\sqrt{\lambda}\,\ell\cos\sqrt{\lambda}\,\ell - \sin\sqrt{\lambda}\,\ell = 0$. In this case
$c_2 = (-\sqrt{\lambda}\cos\sqrt{\lambda}\,\ell)(c_3)$ and it follows that the
eigenfunction ϕ_1 is given by
$\phi_1(x) = \sin(\sqrt{\lambda_1}\, x) - \sqrt{\lambda_1}\, x\cos\sqrt{\lambda_1}\,\ell$ where λ_1 is the
smallest positive solution of the equation
$\sqrt{\lambda}\,\ell = \tan\sqrt{\lambda}\,\ell$. A graphical or numerical estimate of λ_1
reveals that $\lambda_1 \cong (4.49)^2/\ell^2$.

26c. The eigenvalue equation is $2(1-\cos x) = x\sin x$, where
$x = \sqrt{\lambda}\,\ell$. Graphing $f(x) = 2(1-\cos x)$ and $g(x) = x\sin x$ we
see that there is an intersection at $x = 2\pi$. In
addition, it appears there might be an intersection for
$0 < x < 1$. Using a Taylor series representation for $f(x)$
and $g(x)$ about $x = 0$, however, shows there is no
intersection for $0 < x < 1$. Of course $x = 0$ is also an
intersection, which yields $\lambda = 0$, which gives the trivial
solution and hence $\lambda = 0$ is not an eigenvalue.

Section 11.4, Page 612

1. We must first find the eigenvalues and normalized
    eigenfunctions of the associated homogeneous problem
    $y'' + \lambda y = 0$, $y(0) = 0$, $y(1) = 0$. This problem has the
    solutions $\phi_n(x) = k_n \sin n\pi x$, for $\lambda_n = n^2\pi^2$, $n = 1, 2, \dots$.
    Choosing k_n so that $\int_0^1 \phi_n^2 dx = 1$ we find $k_n = \sqrt{2}$. Hence
    the solution of the original nonhomogeneous problem is
    given by $y = \sum\limits_{n=1}^{\infty} b_n \phi_n(x)$, where the coefficients b_n are
    found from Eq.(12), $b_n = c_n/(\lambda_n - 2)$ where c_n is given by
    $c_n = \sqrt{2} \int_0^1 x \sin n\pi x\, dx$ (Eq.9). [Note that the original
    problem can be written as $-y'' = 2y + x$ and therefore
    comparison with Eq.(1) yields $f(x) = x$]. Integrating the
    expression for c_n by parts yields $c_n = \sqrt{2}\,(-1)^{n+1}/n\pi$ and
    thus $y = \sum\limits_{n=1}^{\infty} \dfrac{\sqrt{2}\,(-1)^{n+1}}{(n^2\pi^2-2)\,n\pi}\,\sqrt{2}\sin n\pi x$.

2. From Problem 1 of Section 11.3 we have
    $\phi_n = \sqrt{2}\sin[(2n-1)\pi x/2]$ for $\lambda_n = (2n-1)^2\pi^2/4$ and from
    Problem 7 of that section we have
    $c_n = 4\sqrt{2}\,(-1)^{n-1}/(2n-1)^2\pi^2$. Substituting these values
    into $b_n = c_n/(\lambda_n - 2)$ and $y = \sum\limits_{n=1}^{\infty} b_n \phi_n$ yields the desired
    result.

3. Referring to Problem 3 of Section 11.3 we have
    $y = b_0 + \sum\limits_{n=1}^{\infty} b_n(\sqrt{2}\cos n\pi x)$, where $b_n = c_n/(\lambda_n - 2)$ for

n = 0,1,2,... . The rest of the calculations follow
those of Problem 1.

5. Note that the associated eigenvalue problem is the same
as for Problem 1 and that $|1-2x| = 1-2x$ for $0 \le x \le 1/2$
while $|1-2x| = 2x-1$ for $1/2 \le x \le 1$.

10. Since $\mu = \pi^2$ is an eigenvalue of the corresponding
homogeneous equation, Theorem 11.4.1 tells us that a
solution will exist only if $-(a+x)$ is orthogonal to the
corresponding eigenfunction $\sqrt{2} \sin\pi x$. Thus we require
$\int_0^1 (a+x) \sin\pi x dx = 0$, which yields $a = -1/2$. With $a = -1/2$,
we find that $Y = (x-1/2)/\pi^2$ and $y_c = c\sin\pi x + d\cos\pi x$ by
methods of Chapter 3. Setting $y = y_c + Y$ and choosing d
to satisfy the B.C. we obtain the desired family of
solutions.

11. Note that in this case $\mu = 4\pi^2$ and $\phi_2 = \sqrt{2} \sin 2\pi x$ are the
eigenvalue and eigenfuction respectively of the
corresponding homogeneous equation.

12. In this case a solution will exist only if $-a$ is
orthogonal to $\sqrt{2} \cos\pi x$., that is if $\int_0^1 a\cos\pi x dx = 0$.
Since this condition is valid, a family of solutions
exists.

14. Since $\sum_{n=1}^{\infty} c_n \phi_n(x)$ converges to zero we have $\sum_{n=1}^{\infty} c_n \phi_n(x) = 0$.
Multiplying and integrating as suggested yields
$$\int_0^1 [\sum_{n=1}^{\infty} c_n \phi_n(x)] r(x) \phi_m(x) dx = 0 \text{ or}$$
$$\sum_{n=1}^{\infty} c_n \int_0^1 r(x) \phi_n(x) \phi_m(x) dx = 0. \text{ The integral that multiplies}$$
c_n is just δ_{nm} [Eq. (22) of Section 11.3]. Thus the
infinite sum becomes c_m and the last equation yields
$c_m = 0$.

18. A twice differentiable function v satisfying the boundary
conditions can be found by assuming that $v = ax + b$.
Thus $v(0) = b = 1$ and $v(1) = a + 1$ while $v'(1) = a$.
Hence $2a + 1 = -2$ or $a = -3/2$ and $v(x) = 1 - 3x/2$.
Assuming $y = u + v$ we have $(u+v)'' + 2(u+v) =$

$u'' + 2u + 2(1-3x/2) = 2 - 4x$ or $u'' + 2u = -x$, $u(0) = 0$, $u(1) + u'(1) = 0$ which is the same as Example 1 of the text.

19. From Eq.(30) we assume $u(x,t) = \sum_{n=1}^{\infty} b_n(t)\phi_n(x)$, where the

ϕ_n are the eigenfunctions of the related eigenvalue problem $y'' + \lambda y = 0$, $y(0) = 0$, $y'(1) = 0$ and the $b_n(t)$ are given by Eq.(42). From Problem 1, Section 11.3 we have $\phi_n = \sqrt{2}\sin[(2n-1)\pi x/2]$ and $\lambda_n = (2n-1)^2\pi^2/4$. To evaluate Eq.(42) we need to calculate
$\alpha_n = \int_0^1 \sin(\pi x/2)\sqrt{2}\sin[(2n-1)\pi x/2\,dx$ [Eq.(41) with
$r(x) = 1$ and $f(x) = \sin(\pi x/2)$], which is zero except for $n = 1$ in which case $\alpha_1 = \sqrt{2}/2$, and
$\gamma_n = \int_0^1 (-x)\sqrt{2}\sin[(2n-1)\pi x/2]dx$ [Eq.(35) with
$F(x,t) = -x$]. From Problem 2 we have
$\gamma_n = -4\sqrt{2}(-1)^{n+1}/(2n-1)^2\pi^2 = -c_n$, $n = 1,2,\ldots$. Setting
$\gamma_n = -c_n$ in Eq.(42) we then have
$$b_1 = \frac{\sqrt{2}}{2}e^{-\pi^2 t/4} - c_1\int_0^t e^{-\pi^2(t-s)/4}ds$$
$$= \frac{\sqrt{2}}{2}e^{-\pi^2 t/4} - \frac{4c_1}{\pi^2}e^{-\pi^2(t-s)/4}\Big|_0^t$$
$$= \frac{\sqrt{2}}{2}e^{-\pi^2 t/4} - \frac{4c_1}{\pi^2} + \frac{4c_1}{\pi^2}e^{-\pi^2 t/4} \text{ and}$$

similarly $b_n = -c_n\int_0^t e^{-\lambda_n(t-s)}ds = -(c_n/\lambda_n)e^{-\lambda_n(t-s)}\Big|_0^t$

$$= -(c_n/\lambda_n)(1-e^{-\lambda_n t}), \text{ where } \lambda_n = (2n-1)^2\pi^2/4,$$
$n = 2,3,\ldots$. Substituting these values for b_n along with $\phi_n = \sqrt{2}\sin[(2n-1)\pi x/2]$ into the series for $u(x,t)$ yields the solution to the given problem.

22. In this case $\alpha_n = 0$ for all n and γ_n is given by
$\gamma_n = \int_0^1 e^{-t}(1-x)\sqrt{2}\sin[(2n-1)\pi x/2]dx$
$= e^{-t}\int_0^1 (1-x)\sqrt{2}\sin[(2n-1)\pi x/2]dx$. This last integral
can be written as the sum of two integrals, each of which has been evaluated in either Problem 6 or 7 of Section 11.3. Letting c_n denote the value obtained, we then have

$\gamma_n = c_n e^{-t}$ and thus $b_n = c_n \int_0^t e^{-\lambda_n(t-s)} e^{-s} ds =$

$c_n e^{-\lambda_n t} \int_0^t e^{(\lambda_n - 1)s} ds = [c_n/(\lambda_n - 1)](e^{-t} - e^{-\lambda_n t})$, where

$\lambda_n = (2n-1)^2 \pi^2/4$. Substituting these values into Eq.(30)

yields the desired solution.

24. Using the approach of Problem 23 we find that $v(x)$
satisfies $v'' = 2$, $v(0) = 1$, $v(1) = 0$. Thus
$v(x) = x^2 + c_1 x + c_2$ and the B.C. yield $v(0) = c_2 = 1$ and
$v(1) = 1 + c_1 + 1 = 0$ or $c_1 = -2$. Hence
$v(x) = x^2 - 2x + 1$ and $w(x,t) = u(x,t) - v(x)$ where, from
Problem 23, we have $w_t = w_{xx}$, $w(0,t) = 0$, $w(1,t) = 0$ and
$w(x,0) = x^2 - 2x + 2 - v(x) = 1$. This last problem can
be solved by methods of this section or by methods of
Chapter 10. Using the approach of this section we have

$$w(x,t) = \sum_{n=1}^{\infty} b_n(t) \phi_n(x) \text{ where } \phi_n(x) = \sqrt{2} \sin n\pi x$$

[which are the normalized eigenfunctions of the
associated eigenvalue problem $y'' + \lambda y = 0$, $y(0) = 0$,
$y(1) = 0$] and the b_n are given by Eq.(42). Since the
P.D.E. for $w(x,t)$ is homogeneous Eq.(42) reduces to
$b_n = \alpha_n e^{-\lambda_n t}$ ($\lambda_n = n^2\pi^2$ from the above eigenvalue problem),
where

$\alpha_n = \int_0^1 1 \cdot \sqrt{2} \sin n\pi x dx = \sqrt{2}[1(-1)^n]/n\pi$. Thus

$$u(x,t) = x^2 - 2x + 1 + \sum_{n=1}^{\infty} \frac{\sqrt{2}[1-(-1)^n]}{n\pi} e^{-n^2\pi^2 t} \sqrt{2} \sin n\pi x,$$

which simplifies to the desired solution.

28a. Using variation of parameters we assume
$Y(x) = u_1(x) + xu_2(x)$. Then $Y' = u_2$ since we require
$u_1' + xu_2' = 0$. Differentiating again yields $Y'' = u_2'$ and
thus $u_2' = -f(x)$ by substitution into the D.E. Hence
$u_2(x) = -\int_0^x f(s) ds$, $u_1' = xf(x)$, and $u_1(x) = \int_0^x sf(s) ds$.
Therefore $Y = \int_0^x sf(s) ds - x \int_0^x f(s) ds = -\int_0^x (x-s) f(s) ds$ and
$\phi(x) = c_1 + c_2 x - \int_0^x (x-s) f(s) ds$.

28c. From parts a and b we have

$$\phi(x) = x\int_0^1 (1-s)f(s)\,ds - \int_0^x (x-s)f(s)\,ds$$

$$= \int_0^x x(1-s)f(s)\,ds + \int_x^1 x(1-s)f(s)\,ds - \int_0^x (x-s)f(s)\,ds$$

$$= \int_0^x (x-xs-x+s)f(s)\,ds + \int_x^1 x(1-s)f(s)\,ds$$

$$= \int_0^x s(1-x)f(s)\,ds + \int_x^1 x(1-s)f(s)\,ds.$$

30b. In this case $y_1(x) = \sin x$ and $y_2(x) = \sin(1-x)$ [assume $y_2(x) = c_1\cos x + c_2\sin x$, let $x = 1$, solve for c_2 in terms of c_1 using $y(1) = 0$ and then let $c_1 = \sin 1$]. Using these functions for y_1 and y_2 we find $W(y_1,y_2) = \sin 1$ and thus $G(x,s) = -\sin s\,\sin(1-x)/(-\sin 1)$, since $p(x) = 1$, for $0 \le s \le x$. A similar calculation verifies $G(x,s)$ for $x \le s \le 1$.

30c. Since $W(y_1,y_2)(x) = y_1(x)y_2'(x) - y_2(x)y_1'(x)$ we find that

$$[p(x)W(y_1,y_2)(x)]' = p'(x)[y_1(x)y_2'(x) - y_2(x)y_1'(x)]$$

$$+ p(x)[y_1'(x)y_2'(x) + y_1(x)y_2''(x) - y_2'(x)y_1'(x) - y_2(x)y_1''(x)]$$

$$= y_1[py_2']' - y_2[py_1']' = y_1[q(x)y_2] - y_2[q(x)y_1] = 0.$$

30d. Let $c = p(x)W(y_1,y_2)(x)$. If $0 \le s \le x$, then $G(x,s) = -y_1(s)y_2(x)/c$. Since the first argument in $G(s,x)$ is less than the second argument, the bottom expression of formula (iv) must be used to determine $G(s,x)$. Thus, $G(s,x) = -y_1(s)y_2(x)/c$. A similar argument holds if $x \le s \le 1$.

33. In general $y(x) = c_1\cos x + c_2\sin x$. For $y'(0) = 0$ we must choose $c_2 = 0$ and thus $y_1(x) = \cos x$. For $y(1) = 0$ we have $c_1\cos 1 + c_2\sin 1 = 0$, which yields $c_2 = -c_1(\cos 1)/\sin 1$ and thus $y_2(x) = c_1\cos x - c_1(\cos 1)\sin x/\sin 1$

$$= c_1(\sin 1\cos x - \cos 1\sin x)/\sin 1$$

$$= \sin(1-x) \text{ [by setting } c_1 = \sin 1].$$

Furthermore, $W(y_1,y_2) = -\cos 1$ and thus

$$G(x,s) = \begin{cases} \dfrac{\cos s \, \sin(1-x)}{\cos 1} & 0 \le s \le x \\[3mm] \dfrac{\cos x \, \sin(1-s)}{\cos 1} & x \le s \le 1 \end{cases},$$

and hence $\phi(x) = \int_0^x [\cos s \, \sin(1-x) f(s)/\cos 1] ds +$
$\int_x^1 [\cos x \, \sin(1-s) f(s)/\cos 1] ds$ is the solution of the given
B.V.P.

Section 11.5, Page 623

2a. The general solution of the D.E. is
$y = c_1 J_0(\sqrt{\lambda}\, x) + c_2 Y_0(\sqrt{\lambda}\, x)$ by Eq.(7). The B.C. at $x = 0$
requires that $c_2 = 0$, and the B.C. at $x = 1$ requires
$c_1 \sqrt{\lambda}\, J_0'(\sqrt{\lambda}) = 0$. For $\lambda = 0$ we have $\phi_0(x) = J_0(0) = 1$
and if λ_n is the nth positive root of $J_0'(\sqrt{\lambda}) = 0$ then
$\phi_n(x) = J_0(\sqrt{\lambda_n}\, x)$. Note that for $\lambda = 0$ the D.E. becomes
$(xy')' = 0$, which has the general solution $y = c_1 \ln x + c_2$.
To satisfy the bounded conditions at $x = 0$ we must choose
$c_1 = 0$, thus obtaining the same solution as above.

2b. For $n \ne 0$, set $y = J_0(\sqrt{\lambda_n}\, x)$ in the D.E. and integrate
from 0 to 1 to obtain $-\int_0^1 (xJ_0')' dx = \lambda_n \int_0^1 x J_0(\sqrt{\lambda_n}\, x) dx$.
Integrating the left side of this equation yields
$\int_0^1 (xJ_0')' dx = xJ_0'(\sqrt{\lambda_n}\, x)\Big|_0^1 = J_0'(\sqrt{\lambda_n}) - 0 = 0$ since the λ_n
are eigenvalues from part a. Thus $\int_0^1 x J_0(\sqrt{\lambda_n}\, x) dx = 0$.
For other n and m, we let $L[y] = -(xy')'$. Then
$L[J_0(\sqrt{\lambda_n}\, x)] = \lambda_n x J_0(\sqrt{\lambda_n}\, x)$ and
$L[J_0(\sqrt{\lambda_m}\, x)] = \lambda_m x J_0(\sqrt{\lambda_m}\, x)$. Multiply the first equation
by $J_0(\sqrt{\lambda_m}\, x)$, the second by $J_0(\sqrt{\lambda_n}\, x)$, subtract the
second from the first, and integrate from 0 to 1 to
obtain
$\int_0^1 \{J_0(\sqrt{\lambda_m}\, x) L[J_0(\sqrt{\lambda_n}\, x)] - J_0(\sqrt{\lambda_n}\, x) L[J_0(\sqrt{\lambda_m}\, x)]\} dx =$
$(\lambda_n - \lambda_m) \int_0^1 x J_0(\sqrt{\lambda_n}\, x) J_0(\sqrt{\lambda_m}\, x) dx$. Again the left side
is zero after each term is integrated by parts once, as

was done above. If $\lambda_n \neq \lambda_m$, the result follows with
$\phi_n(x) = J_0(\sqrt{\lambda_n}\, x)$.

2c. We assume that $y = b_0 + \sum_{n=1}^{\infty} b_n J_0(\sqrt{\lambda_n}\, x)$. Since

$-[xJ_0'(\sqrt{\lambda_n}\, x)]' = \lambda_n x J_0(\sqrt{\lambda_n}\, x)$, $n = 0, 1, \ldots$, we find that

$-(xy')' = x \sum_{n=1}^{\infty} \lambda_n b_n J_0(\sqrt{\lambda_n}\, x)$ [note that $\lambda_0 = 0$ and hence b_0
is missing on the right]. Now assume

$f(x)/x = c_0 + \sum_{n=1}^{\infty} c_n J_0(\sqrt{\lambda_n}\, x)$. Multiplying both sides by

$xJ_0(\sqrt{\lambda_m}\, x)$, integrating from 0 to 1 and using the
orthogonality relations of part b, we find
$c_n = \int_0^1 f(x) J_0(\sqrt{\lambda_n}\, x)\, dx / \int_0^1 x J_0^2(\sqrt{\lambda_n}\, x)\, dx$, $n = 0, 1, 2, \ldots$.
[Note that $c_0 = 2\int_0^1 f(x)\, dx$ since the denominator can be
integrated.] Substituting the series for y and $f(x)/x$
into the D.E., using the above result for $-(xy')'$, and
simplifying we find that

$(\mu b_0 + c_0) + \sum_{n=1}^{\infty} [c_n - b_n(\lambda_n - \mu)] J_0(\sqrt{\lambda_n}\, x) = 0$. Thus

$b_0 = -c_0/\mu$ and $b_n = c_n/(\lambda_n - \mu)$, $n = 1, 2, \ldots$, where $\sqrt{\lambda_n}$
are obtained from $J_0'(\sqrt{\lambda_n}) = 0$.

4a. Let $L[y] = -[(1-x^2)y']'$. Then $L[\phi_n] = \lambda_n \phi_n$ and
$L[\phi_m] = \lambda_m \phi_m$. Multiply the first equation by ϕ_m, the
second by ϕ_n, subtract the second from the first, and
integrate from 0 to 1 to obtain
$\int_0^1 (\phi_m L[\phi_n] - \phi_n L[\phi_m])\, dx = (\lambda_n - \lambda_m) \int_0^1 \phi_n \phi_m\, dx$. The integral
on the left side can be shown to be 0 by integrating each
term once by parts. Since $\lambda_n \neq \lambda_m$ if $m \neq n$, the result
follows. Note that the result may also be written as
$\int_0^1 P_{2m-1}(x) P_{2n-1}(x)\, dx = 0$, $m \neq n$.

4b. First let $f(x) = \sum_{n=1}^{\infty} c_n \phi_n(x)$, multiply both sides by $\phi_m(x)$,
and integrate term by term from $x = 0$ to $x = 1$. The
orthogonality condition yields

$c_n = \int_0^1 f(x)\phi_n(x)dx / \int_0^1 \phi_n^2(x)dx$, $n = 1,2,\ldots$ and it is understood that $\phi_n(x) = P_{2n-1}(x)$. Now assume

$y = \sum_{n=1}^{\infty} b_n \phi_n(x)$. As in Problem 2 and in the text

$-[(1-x^2)y']' = \sum_{n=1}^{\infty} \lambda_n b_n \phi_n$ since the ϕ_n are eigenfunctions.

Thus substitution of the series for y and f into the D.E.

and simplification yields $\sum_{n=1}^{\infty} [b_n(\lambda_n - \mu) - c_n]\phi_n(x) = 0$.

Hence $b_n = c_n/(\lambda_n - \mu)$, $n = 1,2,\ldots$ and the desired solution is obtained [after setting $\phi_n(x) = P_{2n-1}(x)$].

Section 11.6, Page 628

1a. Since $u(x,0) = 0$ we have $Y(0) = 0$. However, since the other two boundaries are given by $y = 2x$ and $y = 2(x-2)$ we cannot separate x and y dependence and thus neither X nor Y satisfy homogeneous B.C. at both end points.

2. This problem is very similar to the example worked in the text. The fundamental solutions satisfying the P.D.E.(3), the B.C. $u(1,t) = 0$, $t \geq 0$ and the finiteness condition are given by Eqs.(15) and (16). Thus assume $u(r,t)$ is of the form given by Eq.(17). The I.C. require

that $u(r,0) = \sum_{n=1}^{\infty} c_n J_0(\lambda_n r) = 0$ and

$u_t(r,0) = \sum_{n=1}^{\infty} \lambda_n a k_n J_0(\lambda_n r) = g(r)$. From Eq.(23) of Section

11.5 we obtain $c_n = 0$ and

$\lambda_n k_n a = \int_0^1 rg(r)J_0(\lambda_n r)dr / \int_0^1 rJ_0^2(\lambda_n r)dr$, $n = 1,2,\ldots$.

4. This problem is the same as Problem 17 of Section 10.6. The periodicity condition requires that μ of that problem be an integer and thus substituting $\mu^2 = n^2$ into the previous results yields the given equations.

5a. Substituting $u(r,\theta,z) = R(r)\Theta(\theta)Z(z)$ into Laplace's equation yields $R''\Theta Z + R'\Theta Z/r + R\Theta''Z/r^2 + R\Theta Z'' = 0$ or equivalently $R''/R + R'/rR + \Theta''/r^2\Theta = -Z''/Z = \sigma$. In order

to satisfy arbitrary B.C. it can be shown that σ must be
negative, so assume $\sigma = -\lambda^2$, and thus $Z'' - \lambda^2 Z = 0$ and,
after some algebra, it follows that
$r^2 R''/R + r R'/R + \lambda^2 r^2 = -\Theta''/\Theta = \alpha$. The periodicity
condition $\Theta(0) = \Theta(2\pi)$ requires that $\sqrt{\alpha}$ be an integer n
so $\alpha = n^2$. Thus $r^2 R'' + r R' + (\lambda^2 r^2 - n^2) R = 0$,
$\Theta'' + n^2 \Theta = 0$, and $Z'' - \lambda^2 Z = 0$.

5b. If $u(r, \theta, z)$ is independent of θ, then the $\Theta''/r^2\Theta$ term
does not appear in the second equation of part a and thus
$R''/R + R'/rR = -Z''/Z = -\lambda^2$, from which the desired result
follows.

6. Assuming that $u(r, z) = R(r) Z(z)$ it follows from Problem 5
that $R = c_1 J_0(\lambda r) + c_2 Y_0(\lambda r)$ and $Z = k_1 e^{-\lambda z} + k_2 e^{\lambda z}$. Since
$u(r, z)$ is bounded as $r \to 0$ and $z \to \infty$ we require that
$c_2 = 0$, $k_2 = 0$. The B.C. $u(1, z) = 0$ requires that
$J_0(\lambda) = 0$ leading to an infinite set of discrete positive
eigenvalues $\lambda_1 \lambda_2, \ldots \lambda_n \ldots$. The fundamental solutions of
the problem are then $u_n(r, z) = J_0(\lambda_n r) e^{-\lambda_n z}$, $n = 1, 2, \ldots$.

Thus assume $u(r, z) = \sum_{n=1}^{\infty} c_n J_0(\lambda_n r) e^{-\lambda_n z}$. The B.C.
$u(r, 0) = f(r)$, $0 \le r \le 1$ requires that
$u(r, 0) = \sum_{n=1}^{\infty} c_n J_0(\lambda_n r) = f(r)$ so
$c_n = \int_0^1 r f(r) J_0(\lambda_n r) dr / \int_0^1 r J_0^2(\lambda_n r) dr$, $n = 1, 2, \ldots$.

7b. Again Θ periodic of period 2π implies $\lambda^2 = n^2$. Thus the
solutions to the D.E. are $R(r) = c_1 J_n(kr) + c_2 Y_n(kr)$ and
$\Theta(0) = d_1 \cos n\theta + d_2 \sin n\theta$.

9a. Substituting $u(\rho, \theta, \phi) = P(\rho) \Theta(\theta) \Phi(\phi)$ into Laplace's
equation leads to
$\rho^2 P''/P + 2\rho P'/P = -(\csc^2\phi) \Theta''/\Theta - \Phi''/\Phi - (\cot\phi) \Phi'/\Phi = \sigma$.
In order to satisfy arbitrary B.C. it can be shown that σ
must be positive, so assume $\sigma = \mu^2$.
Thus $\rho^2 P'' + 2\rho P' - \mu^2 P = 0$. Then we have
$(\sin^2\phi) \Phi''/\Phi + (\sin\phi\cos\phi) \Phi'/\Phi + \mu^2 \sin^2\phi = -\Theta''/\Theta = \alpha$.
The periodicity condition $\Theta(0) = \Theta(2\pi)$ requires that $\sqrt{\alpha}$
be an integer λ so $\alpha = \lambda^2$. Hence $\Theta'' + \lambda^2\Theta = 0$ and
$(\sin^2\phi) \Phi'' + (\sin\phi\cos\phi) \Phi' + (\mu^2 \sin^2\phi - \lambda^2) \Phi = 0$.

10. The general solution to the Euler equation is

$P = c_1\rho^{r_1} + c_2\rho^{r_2}$ where $r_1 = (-1+\sqrt{1+4\mu^2})/2 > 0$ and
$r_2 = (-1-\sqrt{1+4\mu^2})/2 < 0$. Since we want u to be bounded
as $\rho \to 0$, we set $c_2 = 0$. As found in Problem 20 of
Section 5.3, the solutions of Legendre's equation are
either singular at 1, at -1, or at both unless
$\mu^2 = n(n+1)$, where n is an integer. In this case, one of
the two linearly independent solutions is a polynomial
denoted by P_n (Problems 21 and 22 of Section 5.3). Since
$r_1 = (-1 + \sqrt{1+4n(n+1)})/2 = n$, the fundamental solutions
of this problem satisfying the finiteness condition are
$u_n(\rho,\phi) = \rho^n P_n(\cos\phi)$, $n = 1,2,\ldots$. It can be shown that
an arbitrary piecewise continuous function on $[-1,1]$ can
be expressed as a linear combination of Legendre
polynomials. Hence we assume that

$u(\rho,\phi) = \sum_{n=1}^{\infty} c_n\rho^n P_n(\cos\phi)$. The B.C. $u(1,\phi) = f(\phi)$ requires

that $u(1,\phi) = \sum_{n=1}^{\infty} c_n P_n(\cos\phi) = f(\phi)$, $0 \le \phi \le \pi$. From

Problem 27 of Section 5.3 we know that $P_n(x)$ are
orthogonal. However here we have $P_n(\cos\phi)$ and thus we
must rewrite the equation in Problem 9b to find
$-[(\sin\phi)\Phi']' = \mu^2(\sin\phi)\Phi$. Thus $P_n(\cos\phi)$ and $P_m(\cos\phi)$ are
orthogonal with weight function $\sin\phi$. Thus we must
multiply the series expansion for $f(\phi)$ by $\sin\phi P_m(\cos\phi)$
and integrate from 0 to π to obtain
$c_m = \int_0^\pi f(\phi)\sin\phi P_m(\cos\phi)d\phi / \int_0^\pi \sin\phi P_m^2(\cos\phi)d\phi$. To obtain
the answer as given in the text let $s = \cos\phi$.

Section 11.7, Page 636

1a. Write $S_n(x)$ as $n\sqrt{x}/e^{nx^2/2}$ and use L'Hopitals Rule.

3. Since $f(x)$ is defined only on the open interval we see
   that $f(x)$ can get as close to 1 as desired, but $f(x)$ is
   never equal to 1. Thus the least upper bound is 1, but
   there is no maximum value.

7.   Expanding the integrand we get

$$R_n = \int_0^1 r(x)[f(x) - S_n(x)]^2 dx = \int_0^1 r(x)f^2(x)dx$$

$$-2\sum_{i=1}^n c_i \int_0^1 r(x)f(x)\phi_1(x)dx + \sum_{i=1}^n \sum_{j=1}^n c_i c_j \int_0^1 r(x)\phi_1(x)\phi_j(x)dx,$$

where the last term is obtained by calculating $S_n^2(x)$.
Using Eqs.(1) and (9) this becomes

$$R_n = \int_0^1 r(x)f^2(x)dx = -2\sum_{i=1}^n c_i a_i + \sum_{i=1}^n c_1^2$$

$$= \int_0^1 r(x)f^2(x)dx - \sum_{i=1}^n a_i^2 + \sum_{i=1}^n (c_1 - a_1)^2, \text{ by completing}$$

the square. Since all terms involve a real quantity
squared (and $r(x) > 0$) we may conclude R_n is minimized by
choosing $c_i = a_i$. This can also be shown by calculating
$\partial R_n/\partial c_i = 2(c_i - a_i)$ and setting equal to zero.

9b. From part a we have $f_0(x) = 1$ and thus $f_1(x) = c_1 + c_2 x$
must satisfy $(f_0, f_1) = \int_0^1 (c_1 + c_2 x)dx = 0$ and
$(f_1, f_1) = \int_0^1 (c_1 + c_2 x)^2 dx = 1$. Evaluating the integrals
yields $c_1 + c_2/2 = 0$ and $c_1^2 + c_1 c_2 + c_2^2/3 = 1$, which have
the solution $c_1 = \sqrt{3}$, $c_2 = -2\sqrt{3}$ and thus
$f_1(x) = \sqrt{3}(1-2x)$.

9c. $f_2(x) = c_1 + c_2 x + c_3 x^2$ must satisfy $(f_0, f_2) = 0$,
$(f_1, f_2) = 0$ and $(f_2, f_2) = 1$.

9d. For $g_2(x) = c_1 + c_2 x + c_3 x^2$ we have $(g_0, g_2) = 0$ and
$(g_1, g_2) = 0$, which yield the same ratio of coefficients
as found in 9c. Thus $g_2(x) = cf_2(x)$, where c may now be
found from $g_2(1) = 1$.

10. This problem follows the pattern of Problem 9 except now
the limits on the orthogonality integral are from -1 to
1. That is $(P_i, P_j) = \int_{-1}^1 P_i(x)P_j(x)dx = 0$, $i \neq j$. For
$i = 0$ and $j = 1$ we have

$$\int_{-1}^1 (c_1, c_2 x)dx = (c_1 x + c_2 x^2/2)\Big|_{-1}^1 = 2c_1 = 0 \text{ and thus } P(1) = 1$$

yields $P_1(x) = x$. The others follow in a similar fashion.

11a. This part has essentially been worked in Problem 7 by setting $c_i = a_i$.

11b. Eq. (6) shows that $R_n \geq 0$ since $r(x) \geq 0$ and thus

$$\int_0^1 r(x) f^2(x) \, dx - \sum_{i=1}^n a_i^2 \geq 0. \quad \text{The result follows.}$$

11c. Since f is square integrable, $\int_0^1 r(x) f^2(x) \, dx = M < \infty$ and therefore the monotone increasing sequence of partial sums $T_n = \sum_{i=1}^n a_i^2$ is bounded above. Thus $\lim_{n \to \infty} T_n$ exists, which proves the convergence of the given sum.

11e. By definition if $\sum_{i=1}^{\infty} a_i \phi_i(x)$ converges to $f(x)$ in the mean, then $R_n \to 0$ as $n \to \infty$. Hence $\int_0^1 r(x) f^2(x) \, dx = \sum_{i=1}^{\infty} a_i^2$.

Conversely, if $\int_0^1 r(x) f^2(x) \, dx = \sum_{i=1}^{\infty} a_i^2$, $\lim_{n \to \infty} R_n = 0$ and

$\sum_{i=1}^{\infty} a_i \phi_i(x)$ converges to $f(x)$ in the mean.

12. Bessel's inequality implies that $\sum_{i=1}^{\infty} a_i^2$ converges and thus the nth term $a_n \to 0$ as $n \to \infty$.

14. If the series were the eigenfunction series for a square integrable function, the series $\sum_{i=1}^{\infty} a_i^2$ would have to converge. But $a_0 = 1$, $a_1 = 1/\sqrt{2}, \ldots, a_n = 1/\sqrt{n}, \ldots,$ and $\sum_{n=1}^{\infty} a_n^2 = \sum_{n=1}^{\infty} 1/n$ is the well-known harmonic series which does not converge.